The Chemistry of Aromatherapeutic Oils

The Chemistry of Aromatherapeutic Oils

3rd Edition

E. Joy Bowles

Routledge
Taylor & Francis Group

LONDON AND NEW YORK

First published 2003 by Allen & Unwin

Published 2020 by Routledge
2 Park Square, Milton Park, Abingdon, Oxon OX14 4RN
605 Third Avenue, New York, NY 10017

Routledge is an imprint of the Taylor & Francis Group, an informa business

National Library of Australia
Cataloguing-in-Publication entry:

Bowles, E. Joy.
The chemistry of aromatherapeutic oils.
Includes index.

1. Essences and essential oils—Therapeutic use. 2. Aromatherapy.
I. Title

Set in 11/13pt Sabon by Midland Typesetters, Victoria

ISBN-13: 9781741140514 (pbk)

CONTENTS

LIST OF ILLUSTRATIONS

LIST OF TABLES

ACKNOWLEDGMENTS

This book is the product of eleven years of research, teaching and learning. As with all creative labours, its writing has required much energy, grace and generosity from my teachers, mentors, colleagues, students, friends and family. For everyone's gifts of inspiration over time I am deeply thankful.

I hesitate to name people individually, as I may forget some, but I do wish to credit Dr D. Pénoël and Pierre Franchomme, whose seminal work *L'aromathérapie exactement* started me off on my journey. I also acknowledge my publisher Emma Cant and the team at Allen & Unwin who have made my transition from self-publishing to commercially published author a pleasant and easy one. I am indebted to my reviewers including Hans Wohlmuth, David de Vries, Corinne Border and Bob Harris who made valuable comments and corrections in the editing process.

It is my hope that this book will aid communication between aromatherapists and other health professionals, and contribute to an understanding of the therapeutic principles of the essential oils.

E. Joy Bowles, Lismore, Australia
January 2003

FOREWORD

A recent poll of scientists' opinions about alternative medicine revealed a perhaps surprising general level of acceptance, but this was biased in favour of 'serious' therapies such as acupuncture and osteopathy. Aromatherapy was not given much credence, although it was awarded some credit for 'relaxation'.

The popular perception of aromatherapy is pretty far removed from the kind of aromatherapy being taught, practised and written about in books such as this. It is not really surprising that the lack of knowledge about, and understanding of, aromatherapy is widespread, and shared even by scientists. The volume of clinical research concerning essential oils is small, almost all of it has been published since 1990, and very few books even refer to it.

We need more books that discuss the scientific literature which, if you step beyond the clinical realm, is bordering on the substantial, especially in relation to certain essential oils and aromachemicals. Any study of aromatherapy and essential oils is hollow, unless the basic chemistry is understood, and this text does an excellent job of imparting that knowledge.

This is not a new text, but is a substantial and very welcome claboration of the previous edition. The new edition has a depth and breadth through which the writer dovetails chemistry with pharmacology and toxicology. It is no secret that I am not a proponent of the idea that all the chemical constituents in a particular category share certain properties, and I am pleased to find that Joy has taken a more balanced approach in this new edition. At the same time it is always fascinating to discover that a certain property is explainable because of some molecular foible, and Joy does an excellent job of joining up the knots.

It is monumentally difficult to write about chemistry when most of your target audience finds it an unfamiliar and obtuse subject, but Joy manages to convey her enthusiasm and insight in a way that both educates and inspires.

Robert Tisserand
California, September 2003

INTRODUCTION

People in today's highly pressurised society are increasingly being affected by new forms of disease that are less amenable to treatment with mainstream medicine. Doctors and specialists seem to have less and less time to spend on consultations, and are often perceived as being too readily influenced by the pharmaceutical industry's incentives.

Many people disappointed with mainstream treatments turn to complementary therapies, finding there a different approach to health and disease. Most complementary therapies, however, rely on historical and empirical evidence rather than scientific evidence. The scientific method uses randomised controlled trials to evaluate efficacy of a drug or therapy. Randomisation aims to reduce any bias in selection of participants, and using a control group of people with the same characteristics as your treatment group allows you to compare the effect of a treatment with a placebo, or with another treatment. A placebo or non-active treatment is used to control for the effect that people's beliefs about the treatment being tested may have on the outcome of the treatment.

Scientific evaluation of aromatherapy faces some difficulties: finding a fragrant placebo that has no effect; assessing the effect of the consultation and practitioner–client relationship; and separating olfactory effects from pharmacological effects and the effects of massage. However until there have been enough randomised clinical trials, there are many studies of essential oils and their components that can be used to hypothesise the likely effects of essential oils in humans.

Why study essential oil chemistry?

The purpose of this book is to offer you a way of understanding essential oil chemistry so that you can make sense of the research literature. A knowledge of essential oil chemistry also enables you to:

- make 'educated guesses' about the properties of unfamiliar oils once you can identify their chemical composition. Let's take the Australian oil Rosalina, *Melaleuca ericifolia*, as an example. A quick search on the Internet yields several sites that list one of its major components as linalool. Given this as a starting point, you can research the known effects of linalool by a search of the Medline database, and begin to get a feel for how the oil may be used. If you knew more of the constituents, you could begin to build up a composite profile of possible effects of the whole oil.[1]
- make judgements about the safety of different oils. If an adverse event occurs while using an essential oil, you can look at the hazards or toxicity associated with its individual constituents, and estimate whether any were likely to have caused the event.

What does this book cover?

The book starts with a brief overview of chemistry and a simplified explanation of how atoms bond to form the molecules found in essential oils. Chapter 2 describes the aspects of plant anatomy and physiology that are relevant to essential oils, and the four main types of molecule found in essential oils: terpenoids, phenols and phenyl propanoids, non-terpenoid aliphatic molecules and heterocyclic compounds. Chapter 3 discusses terpenes and the polarity and solubility of essential oil molecules. Chapter 4 illustrates the different functional groups found in essential oil molecules with examples of each from the research literature.

Chapter 5 offers a brief overview of pharmacology and then explores possible actions of essential oils on the human body given the information in Chapters 3 and 4. Chapter 6 is concerned with quality control, Chapter 7 with some more advanced details about the chemical naming system for essential oil molecules. Chapter 8 has a 'Ready Reference' list of the top 3 constituents found in 94 different essential oils, and chemical structures for several of the common constituents from each functional group.

1

WHAT IS CHEMISTRY?

Chemistry is the study of the composition, properties and transformations of substances. One has to distinguish between physical properties (shape, colour, texture, etc.) and chemical properties such as the fundamental transformations a substance, for example, wood, undergoes when tested.

One of the simplest transformational experiments is to see what happens when they are heated. If you heat a piece of wood sufficiently, it will catch alight and burn with a yellow flame, producing smoke and some grey or black ash. If you heat a piece of glass sufficiently, it may melt and change shape, but it won't burn with a yellow flame or produce smoke and ash.

Chemists have systematically studied thousands of different types of materials using many different types of transformational experiments. The result of their research is the discovery that there are about 105 naturally occurring fundamental substances, known as elements. Some of these elements occur as everyday materials, but most elements more usually combine with other elements to form compounds. Examples of elemental and compound substances are shown in Table 1.1. The formulae are a chemistry short-hand showing the types of atoms present in the substances.

Table 1.1   Examples of elements and compounds

Elements			Compounds		
Substance name	Formula	Type of atom	Substance name	Formula	Types of atom
Hydrogen gas	H_2	Hydrogen	Water	H_2O	Hydrogen, oxygen
Oxygen gas	O_2	Oxygen	Carbon dioxide	CO_2	Carbon, oxygen
Diamond	C	Carbon	Butane	C_4H_{10}	Carbon, hydrogen
Sulphur	S	Sulphur	Hydrogen sulphide (rotten egg gas)	H_2S	Hydrogen, sulphur
Silver metal	Ag	Silver	Silver oxide (black tarnish on silver)	AgO	Silver, oxygen
Nitrogen gas	N_2	Nitrogen	Ammonia	NH_3	Nitrogen, hydrogen

Atomic theory of elements

To explain how elements combined to form compounds with very different properties than those of the component elements, a theory of atoms was gradually developed during the 1800s.[1] The theory goes like this:

- Elements are made up of indivisible particles called atoms. Atoms of a given element all share the same properties.
- Chemical changes occur when atoms are combined, separated or rearranged.
- Atoms of different elements have different properties.
- Atoms combine together in fixed whole number ratios to form larger particles known as molecules.

In 1909, Ernest Rutherford proposed that atoms were mainly empty space, with a tiny, positively charged central nucleus orbited by a series of negatively charged electrons. This particle theory of atoms has been superseded by quantum mechanics, but it is still useful for explaining how different elements combine to form molecules.

Rutherford's atomic model

In Rutherford's model of the atom, atoms are made up of three types of subatomic particles of matter called protons, neutrons and electrons. Each subatomic particle can be thought of as a building block, but single protons, neutrons and electrons do not manifest the characteristics of any element. It is the specific number and ratio of protons, neutrons and electrons that together give rise to atoms of different elements.

Protons and electrons are electrically charged. Positively charged protons attract negatively charged electrons, but repel other protons. Electrons, being negatively charged, repel other electrons. The rule is 'opposite charges attract, similar charges repel'. Neutrons, as the name suggests, are neutral: they have no charge and are not affected by the charge of protons or electrons. In Rutherford's model, protons, neutrons and electrons are arranged within atoms in the following ways:

- The number of protons equals the number of electrons, so the total charge on the atom, when they are added together, is zero.
- The protons are in the centre or nucleus of the atom.
- The neutrons are in the nucleus and seem to prevent the positive charges from repelling each other.
- The electrons orbit the nucleus at defined energy levels or orbitals.

Imagine an electron in an orbital as a yoyo being whirled on its string. Most of the time, in any one spot on the circumference of the yoyo's path, there is empty space, but if you put your hand in the way of the whirling yoyo, it does not feel like empty space. Figure 1.1 shows two ways of representing atomic structure. The dots are the subatomic particles.

Figure 1.1   Two ways to represent atomic structure

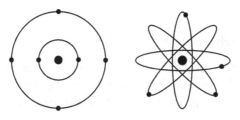

Types of atoms found in essential oils

Like most molecules made by living organisms, essential oil molecules are mainly made up of hydrogen and carbon atoms. Several types of essential oil molecule also contain oxygen atoms, though these are usually in the functional groups attached to the carbon skeleton of the molecule (see Table 1.4 'Functional groups' on page 18).

Some essential oil molecules also contain sulphur atoms (for example, diallyl sulphide, from garlic and the onion family), and some contain nitrogen atoms (for example, indole, from Jasmine, and methyl N-methyl anthranilate, from Mandarin and Sweet orange peel oils). We will now look at the structures of hydrogen, carbon and oxygen atoms in more detail.

Structure of a hydrogen atom

Hydrogen is the smallest atom. It has only one proton orbited by one electron. It is unique among atoms in that it doesn't have any neutrons in its nucleus. The electron is located in the first orbital at a set distance from the proton. This first orbital has the potential to contain two electrons, so there is a vacancy in the first orbital of hydrogen atoms. The vacancy determines the capacity of hydrogen atoms to form molecules. Figure 1.2 is a two-dimensional diagram of a hydrogen atom.

Figure 1.2 A hydrogen atom

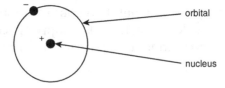

Structure of a carbon atom

The next atom we concern ourselves with is carbon. It has six protons and six electrons. There are also six neutrons in the nucleus of a carbon atom which act as a kind of 'bubble wrap' or insulating

Summary Box 1.1   Hydrogen

Element	Hydrogen
Symbol	H
Number of protons (+)	1
Number of neutrons	0
Number of electrons (-)	1 in first orbital
Number of electron vacancies	1 in first orbital
Number of bonds possible	1

layer to prevent the six positively charged protons from repelling each other. Two electrons fill the first orbital. Four electrons are located in the second orbital. The electrons in the second orbital have more energy than those in the first. They are depicted as being further away from the nucleus. Electrons can jump between orbitals if they are sufficiently energised, which is where the term 'quantum leap' comes from, as there is no continuum between orbitals. The second orbital can contain as many as eight electrons, so there are four vacancies in the second orbital of a carbon atom.

Figure 1.3 below shows the structure of a carbon atom, with its two electron shells.

Figure 1.3   A carbon atom

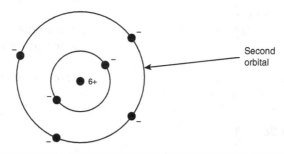

Summary Box 1.2   Carbon

Element	Carbon
Symbol	C
Number of protons (+)	6
Number of neutrons	6
Number of electrons (-)	2 in first orbital
	4 in second orbital
Number of electron vacancies	4 in second orbital
Number of bonds possible	4

Structure of an oxygen atom

An oxygen atom has eight protons and eight neutrons in the nucleus, and also eight electrons. As with the carbon atom, the first electron orbital is full, containing two electrons. The second orbital contains six electrons, leaving two vacancies. Figure 1.4 shows the structure of an oxygen atom.

Figure 1.4   An oxygen atom

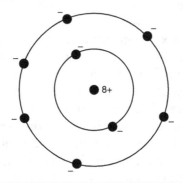

Summary Box 1.3 Oxygen	
Element	**Oxygen**
Symbol	O
Number of protons (+)	8
Number of electrons (-)	2 in first orbital
	6 in second orbital
Number of neutrons	8
Number of electron vacancies	2 in second orbital
Number of bonds possible	2

Information Box 1.1
Ions

Ions are atoms that have either gained or lost one or more electrons. For example, free radical cations (positively charged) are formed when an atom is bombarded by sufficient energy to allow an electron to escape from the atom altogether. Anions are negatively charged because they receive an extra electron.

Ionic bonding occurs when electrons are transferred between ions of opposite charges so that their outermost orbitals are full. Metallic elements and non-metals bind ionically to form substances known as salts. Salts usually occur in a crystalline form, where many positive and negative ions form a lattice held together by electrical charge. Ionic compounds do not exist as single molecules in the solid state.

An example of an ionically bonded substance is common table salt, sodium chloride (NaCl). When dissolved in water, the positively charged sodium ions (Na^+) and negatively charged chlorine ions (Cl^-) dissociate from each other, reforming into the crystal lattice only when the water molecules evaporate. Bodily fluids have many different types of ions dissolved in them, and the right balance of positive and negative ions in body fluids is crucial for cell function.

Bonding

As atoms are neutrally charged, you would not expect them to attract or repel each other. However, most atoms do not exist as single atoms. They bond together with other atoms to form molecules. Hydrogen, carbon and oxygen atoms all have electron vacancies in their outermost orbital. When atoms have electron vacancies, they bind with other atoms to get electrons to fill the vacancies. This is because a full orbital is more stable energetically than a partially full one.

Atoms that don't bond

Helium and neon are examples of substances which do exist as single atoms. A helium atom has two protons, two neutrons and two electrons. This means that its only electron orbital is full, so it has no need to bond with any other atom. A neon atom has ten protons, ten neutrons and ten electrons, which means that both its first and second orbitals are full. You can imagine helium and neon as 'self-actualised' atoms that don't need other atoms to 'feel fulfilled'. Neither helium or neon atoms have yet been discovered in molecules, so they are considered 'inert' or un-reactive elements.

Types of bonding

Bonding is an interaction between two atoms. There are two main types of bonding: ionic and covalent. Information Box 1.1 describes ions and ionic bonding, but as essential oils are covalently bonded, we will not consider ionic bonding further here.

Covalent bonding occurs between atoms by sharing electrons. In order for this to happen, atoms have to get close enough to overlap their outer electron orbitals. Hydrogen gas consists of molecules of hydrogen, each molecule containing two hydrogen atoms bonded together. A hydrogen molecule is the simplest covalently bonded molecule. Figure 1.5 shows two ways of representing a hydrogen atom. The first demonstrates the overlap of orbitals, and how each electron still belongs to its own atom while being accessible to the other atom. The second drawing uses a 'stick' to show the bond between the two atoms, with an electron at either end of the stick.

Figure 1.5   Two ways of representing a hydrogen molecule

Figure 1.6 shows two ways of drawing a molecule of water. Notice how the oxygen atom requires two electrons to fill its outer orbital (eight electrons), and in this case has bonded with two different hydrogen atoms to get them. Notice also that each hydrogen atom now has acquired an extra electron and has a full orbital.

Figure 1.6   Two ways of representing a water molecule

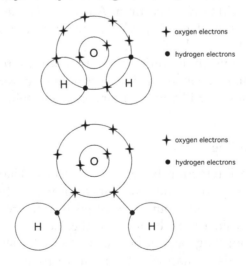

Table 1.2 Numbers of electrons involved in bonding

Atom	No. of electrons in outermost orbital	Maximum possible no. of electrons in outer orbital	Bonds needed for molecular stability
Hydrogen	1	2	1
Carbon	4	8	4
Oxygen	6	8	2

Table 1.2 lists the number of bonds that hydrogen, carbon and oxygen atoms need to make with other atoms in order to gain full outer orbitals.

Double and triple covalent bonds can be formed between two atoms if there are enough electrons to share. A double bond is where two pairs of electrons are shared between the two atoms. Carbon atoms quite often form double bonds with other carbon atoms, as will be seen in the examples of essential oil molecules in Chapters 3 and 4. Figure 1.7 shows two ways of drawing a molecule of ethene, which has two carbon atoms and four hydrogen atoms. Ethene has a double bond between the two carbon atoms. Black dots represent electrons from carbon atoms, diamonds electrons from hydrogen atoms. The inner orbital electrons on the carbon atoms are not shown.

The types of bonds between atoms in a molecule affect its three-dimensional structure. This in turn affects its chemical properties and, as we will see in Chapters 3 and 4, it also affects the odour and therapeutic properties of essential oil molecules.

Symbols, formulae and chemical drawings

When written down, chemistry is a curious mixture of letters and symbols, which allows for a two-dimensional representation of three-dimensional molecular structures, and also provides a short-hand for ease of communication. Atoms are usually represented by the initial of their Latin or Greek name, or first two letters if more than two elements share the same initial. For example, Au is the symbol for gold, from the Latin *aurum*, 'gold'.

Figures 1.1 to 1.7 used dots for subatomic particles and curved lines for electron orbitals, but for large molecules this drawing

Figure 1.7   Two ways of representing an ethene molecule, C₂H₄

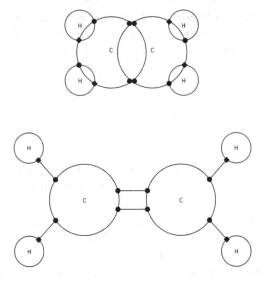

method would get very tedious (and take up far too much space). The simplest way to represent a molecule is by drawing sticks or bars between the atoms, each stick representing one bond.

Water molecules (one atom of oxygen, two atoms of hydrogen) can therefore be represented as in Figure 1.8.

Figure 1.8   A water molecule

Chemical formulae

When chemists want to refer to the atomic components of a molecule, they use its formula. This is an even shorter way of expressing types and numbers of atoms present. For example, water is written H_2O, and ethene C_2H_4. The subscript numbers indicate the number of atoms of each type present (for example, H_4 means there are four atoms of hydrogen present).

For more complex molecules, there are different ways of writing formulae. Let us take ethanol, the substance found in wine referred to as 'alcohol' by the general public. Figure 1.9 shows the arrangement of the two carbon atoms, six hydrogen atoms and one oxygen atom that make up an ethanol molecule.

Figure 1.9 An ethanol molecule

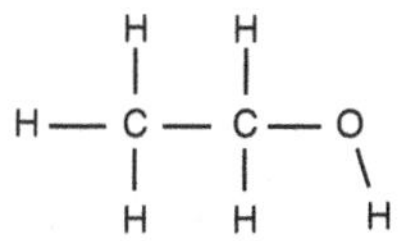

According to convention, in the simplest written formulae, carbon is mentioned first, hydrogen second, and oxygen last. Thus ethanol is written as C_2H_6O. The formula can also be written to give more information about the molecule's structure, for example, CH_3–CH_2–OH. If you look at the stick structure in Figure 1.9, you can see how the carbon atoms form the skeleton of the molecule. Yet another way of writing the formula highlights the placement of oxygen atoms in the molecule, for example, C_2H_5OH.

Drawing chemical structures

Chemists are also interested in representing the physical structure of molecules more closely. In three dimensions, the molecules are not in straight lines as implied in Figure 1.8. The atoms are joined to each other at angles to ensure their maximum separation. This is because the outer electrons round each atom repel electrons round the other atoms.

Thus, if a carbon atom has four other atoms (say, hydrogen atoms) joined to it, the three-dimensional shape of the molecule will be tetrahedral, as in the molecule methane (CH_4) shown in Figure 1.10. Imagine the solid black arrowhead coming out of the page surface and the dashed arrowhead going back behind the page. The diagram next to it in the shape of a cross is how the molecule is represented with just straight lines.

Figure 1.10 Two ways of representing a methane molecule

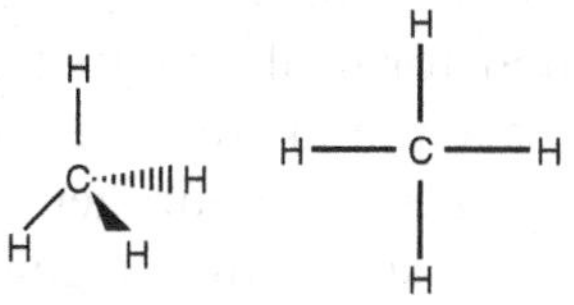

At the site of a double bond, the molecule will have a planar triangular shape like ethene (C_2H_4), as shown in Figure 1.11. Remember that hydrogen cannot make double bonds as it requires only one

Figure 1.11   An ethene molecule   Figure 1.12   An ethyne molecule

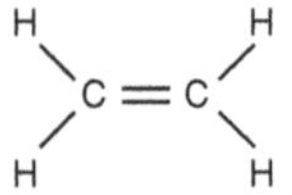

$$
\begin{array}{c}
H \\
\diagdown \\
\diagup C = C \diagdown \\
H H
\end{array}
\qquad\qquad
H{-}C{\equiv}C{-}H
$$

additional electron to achieve a full outer orbital. In essential oils, therefore, only carbon and oxygen atoms will make double bonds.

Two carbon atoms can even share three bonds with each other, in what is called a triple bond. The net result is a linearly shaped molecule known as ethyne (C_2H_2), as shown in Figure 1.12. Triple bonds are not very stable physically, which makes them reactive. Carbon-carbon triple bonds are very uncommon in essential oils.

Carbon atoms are capable of forming many complex structures, due to their ability to form four bonds. Figure 1.13 shows isobutane (C_4H_{10}) as a branched structure and cyclohexane (C_6H_{12}) as a closed ring.

Figure 1.13   Variability in structure of carbon-based molecules

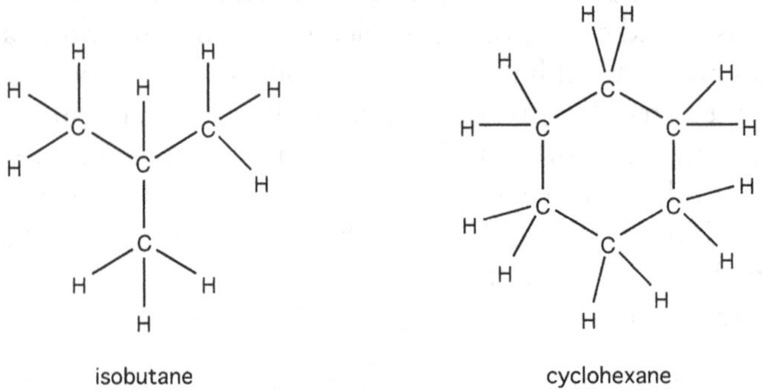

isobutane   cyclohexane

Most essential oil molecules have long carbon chains and require an even more abbreviated form of representation. In the first illustration in Figure 1.14, each stick represents a bond between two carbon atoms. The attachment of hydrogen atoms is presumed and not shown, whereas the position of oxygen atoms is shown. The molecule is linalool, a constituent of Lavender oil and many other essential oils.

Phenolic and coumarin essential oil molecules feature structures known as 'benzene rings'. The molecule C_6H_6 is the compound

Figure 1.14   Two ways of representing a linalool (C₁₀H₁₇OH) molecule, showing ease of the simplified stick method

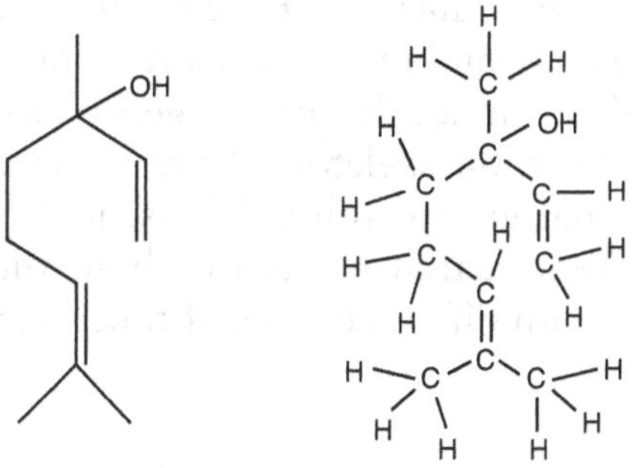

benzene. Its structure is similar to the closed 6-carbon ring of cyclo-hexane (see Figure 1.13), but the bonds are neither single or double. As shown in Figure 1.15, the fourth electron of each carbon atom in a benzene ring appears to be equally shared by the carbon atoms on either side, causing an even spread or 'delocalisation' of electrons over the ring. See 'Phenols' in Chapter 4 for more details.

Figure 1.15   A benzene ring (C₆H₆), showing the way electrons are shared between carbon atoms in the ring

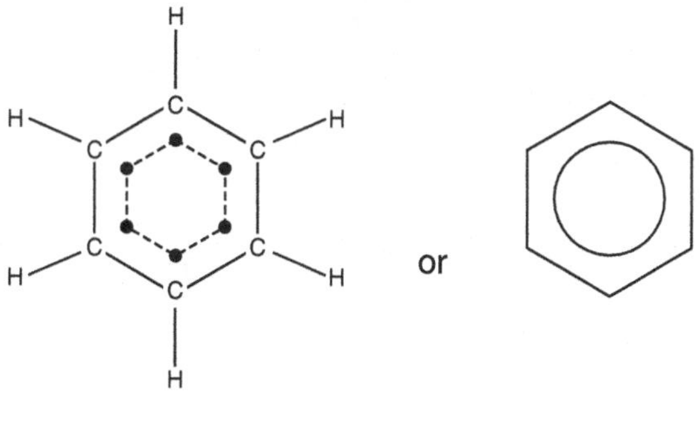

or

Chemical names

The names of carbon-based molecules convey useful information about their structure. Table 1.3 is a key to the meanings of some chemical names. Some of them, such as methane (a greenhouse gas), propane and butane (BBQ and cigarette lighter fuels), may already be familiar.

Functional group names

A functional group is a chemical entity that gives a molecule its particular characteristics, or its 'function'. Functional groups can affect the odour, solubility, toxicity and therapeutic properties of molecules. A number of essential oil molecules have one or more functional groups attached to the carbon skeleton of the molecule. Table 1.4 summarises the functional groups found in essential oil molecules. Chapter 4 considers the research literature about the therapeutic properties of molecules with different types of functional groups.

Further reading

- For an interactive periodic table of elements outlining the characteristics of each chemical try http://www.webelements.com/
- P. Strathern (2001), *Mendeleyev's Dream: The Quest for the Elements*, Penguin Putnam Inc, Berkley, describes the history of chemistry from ancient Greek times in an accessible and, at times, humorous way.

Table 1.3 Simple carbon compounds

Name and formula	Structure	Part of name	Meaning
Methane CH_4		Meth-	One carbon atom
Ethane C_2H_6		Eth-	Two carbon atoms
Propane C_3H_8		Prop-	Three carbon atoms in a chain
Butane C_4H_{10}		But-	Four carbon atoms in a chain
bonds		-ane	Only single between the carbon atoms
Ethene C_2H_4		-ene	A double bond present between two carbon atoms

Table 1.4 Functional groups found in essential oil molecules

Molecule description	Functional group	Name ending	Structure	Location
Alcohol	Hydroxyl	-ol	-OH	Primary = at end of chain; secondary = on C joined to two other Cs; tertiary = on C joined to three other Cs.
Aldehyde	Carbonyl	-al; -aldehyde		At end of C chain
Ketone	Carbonyl	-one		In middle of C chain
Acid	Carboxyl	-ic acid		At end of C chain
Ester	Ester	-yl + -ate		
Phenol	Phenol	-ol		
Ether	Ether	-ole	C-O-C	
Phenyl methyl ether	Phenyl methyl ether	-ole		
Furanoid	Furan	-furan-		O atom part of 5-membered ring

Pyranoid	Pyran	-pyran-		O atom part of 6-membered ring
Oxide	Cyclic ether	-ole		O atom part of closed ring (number of C atoms varies)
Lactone	Lactone	-one; -in		O atom included in closed ring (number of C atoms varies)
Coumarin	Benzo-alpha-pyrone	-in; -en; -one		Benzene ring + lactone ring

2

PLANTS AND ESSENTIAL OILS

Plant classification

In order to understand how plants make essential oils, it is probably helpful to take a step back from the molecular level and look at plants as living organisms. The study of plants is known as botany. Botany uses a taxonomic classification system to arrange the huge variety of plants into groups that share similar anatomical characteristics.

The different groups can be arranged in a branching manner like a family tree, showing the degree of relatedness between different plant families. The names start at the top with Kingdom, Sub-kingdom, Division, Class, Order, Family, Genus, Species and Sub-species. Figure 2.1 shows a diagram of how the levels are arranged. There are 4 sub-kingdoms, 5 superdivisions, 5 divisions, 1 class, 2 orders, 6 families, 7 genera and 50 species of *Pinus*, so not all the groups are shown in the diagram.

There are at least four classification systems in use, but the differences between them occur mainly at higher levels of classification and do not affect the genus and species names too much. The system used in Figure 2.1 is the Cronquist system (see 'Further reading' for references to other systems). Table 2.1 shows common plant families and examples of plant species in each that yield commercially available essential oils.

How to write botanic names

Botanic names are generally derived from Latin or Greek words, as these were the academic languages when taxonomic systems were first invented. The eighteenth-century Swedish naturalist Linnaeus (or Linné) was the first to use two names (the Latin binomial system) to

Figure 2.1　Taxonomic levels used in botany, with example of names for Scotch pine (*Pinus sylvestris*). Not all groups are shown for each level.

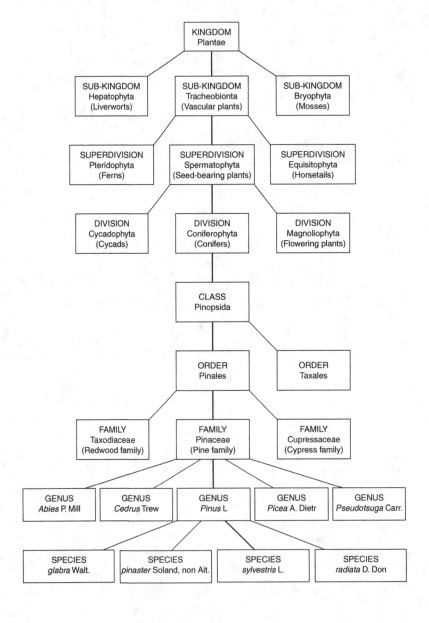

Source: Adapted from United States Department of Agriculture, Plants Database, http://plants.usda.gov/ [accessed 16-05-2003].

classify plants and animals. A large number of genera and species have a capital 'L.' following their names, indicating that it was Linnaeus who first classified them. As new species are discovered, or old ones re-classified, the name of the botanist who most recently created a name gets added in abbreviated form.

The most commonly used botanic names of any specimen are the last two, the genus and the species. They are normally written in italics, with only the genus name having a capital letter, for example *Pinus sylvestris*. A small 'x' between genus and species indicates a hybrid, for example *Mentha* x *piperita*. Above the genus level, the names are written in normal type. Sub-species are minor variations of a species, which differ in colour or structure of a plant part.

Another addition to the botanical name that often occurs on bottles of essential oils is the *chemotype*. Some plants from anatomically identical species produce essential oils of different odour and composition (see 'Variation in essential oil composition' on page 34). The chemotype name usually appears after the italic genus-species binomial: *Rosmarinus officinalis* CT camphor (CT, chemotype; camphor, the distinctive compound in this chemotype).

In an ideal world, all essential oils should be extracted from only one species or one chemotype of a species, from one geographical location and labelled with their botanic name and chemotype. However, the multi-level nature of the industry tends to preclude such stringent quality assurance. Suppliers that source their oils directly from the growers are more likely to be able to determine the exact botanical species of their oils. Chemotypic variation can usually be detected by odour, but should preferably be confirmed by GC–MS analysis (see Chapter 6).

Plant anatomy

Plant anatomy is the study of the physical features of plants, such as leaves, flowers, fruits, shape and size of the plant. Differences in plant anatomical features form the basis for plant classification. To determine which family a plant belongs to, you need to examine the shapes, orientation, structure and numbers of its anatomical features. Variations in leaf width and length, presence or absence of hairs and different flower colours give the genus of a plant. Figure 2.2 shows some of these features as they appear in a sage plant (*Salvia officinalis*).

Figure 2.2   Anatomical structures of *Salvia officinalis* (family Lamiaceae)

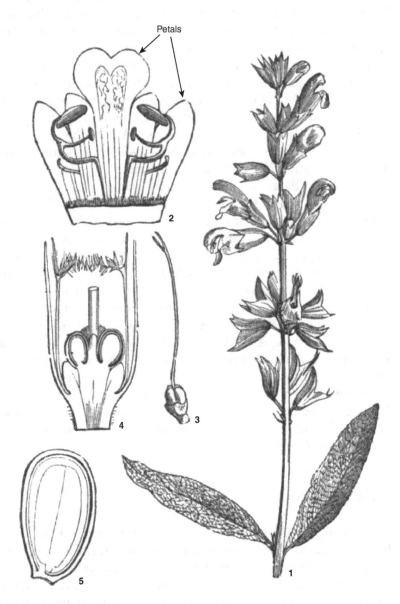

1. *Salvia officinalis*;
2. its corolla laid open;
3. its pistil;
4. the pistil and lower part of the flower cut open;
5. perpendicular section of a nut.

Table 2.1   Common plant families and examples of essential oil producing plants from each family

Plant family	Plant part extracted for essential oil	Common name	Botanic name
Annonaceae	Flowers	Ylang Ylang	*Cananga odorata*
Apiaceae (also known as Umbelliferae)	Fruits and roots	Angelica	*Angelica archangelica*
	Fruits	Aniseed	*Pimpinella anisum*
	Fruits and leaves	Dill	*Anethum graveolens*
	Fruits	Sweet Fennel	*Foeniculum vulgare* var. *dulce*
Asteraceae (also known as Compositae)	Flowering tops	German Chamomile	*Chamomilla recutita*
	Flowering tops	Roman Chamomile	*Anthemis nobilis*
Burseraceae	Resin	Frankincense	*Boswellia carterii*
	Resin	Myrrh	*Commiphora myrrha*
	Resin	Elemi	*Canarium luzonicum*
Cupressaceae	Needles	Cypress	*Cupressus sempervirens*
	Berries and leaves	Juniper	*Juniperus communis*
Geraniaceae	Leaves	Geranium	*Pelargonium graveolens*
Lamiaceae	Leaves and flowering tops	Basil	*Ocimum basilicum*
	Leaves and flowering tops	Clary Sage	*Salvia sclarea*
	Leaves and flowering tops	Lavender	*Lavandula angustifolia*
	Leaves	Oregano	*Origanum vulgare*
	Leaves	Peppermint	*Mentha* x *piperita*
	Leaves	Rosemary	*Rosmarinus officinalis*

Table 2.1 *continued*

Lamiaceae	Leaves	Patchouli	*Pogostemon cablin*
Lauraceae	Heartwood	Rosewood	*Aniba rosaeodora*
	Wood, bark and leaves	Camphor	*Cinnamomum camphora*
	Bark and leaves	Cinnamon	*Cinnamomum zeylanicum*
	Leaves	Bay	*Laurus nobilis*
Myristicaceae	Nuts	Nutmeg	*Myristica fragrans*
Myrtaceae	Leaves	Eucalyptus	*Eucalyptus* sp.
	Leaves	Myrtle	*Myrtus communis*
	Leaves	Tea-tree	*Melaleuca* sp.
	Leaves and dried buds	Clove	*Syzygium aromaticum*
Oleaceae	Flowers	Jasmine	*Jasminum* sp.
Piperaceae	Dried fruits	Black Pepper	*Piper nigrum*
Pinaceae	Needles	Pine	*Pinus* sp.
	Wood	Cedarwood	*Cedrus* sp., *Juniperus* sp.
Poaceae	Grass leaves	Lemongrass, Citronella, Palmarosa	*Cymbopogon* sp.
	Roots	Vetiver	*Vetivera zizanoides*
Rosaceae	Flowers	Rose	*Rosa damascena*
Rutaceae	Fruit peel	Lemon, Mandarin, Grapefruit, Lime, Orange, Bergamot	*Citrus* sp.
	Flowers	Neroli	*Citrus aurantium* var. *amara*
	Leaves and stems	Petitgrain	*Citrus aurantium* var. *amara*
Styraceae	Resin	Benzoin	*Styrax benzoin*
Zingiberaceae	Rhizome	Ginger	*Zingiber officinale*
Santalaceae	Heart wood	Sandalwood	*Santalum* sp.

Essential oil storage structures

Plants that produce essential oils usually store their essential oils in special structures. Examples of these include secretory hairs, secretory cells within the epidermis (outer layer of cells), special sacs made from several secretory cells surrounding an oil-filled space and secretory ducts (tubes lined with secretory cells). The secretory hairs point outwards from the surface of the leaves and stems, quickly releasing their essential oil when the surface is brushed. Oils produced in secretory hairs are usually designed to repel would-be predators.

Secretory sacs and ducts are more often located inside the leaves or in the heartwood or roots. The essential oils in these structures possibly protect the plant against bacteria, fungi and pests like termites. Svoboda et al. (2000) have collected some excellent microscopic photographs of secretory essential oil structures (see 'Further reading').

The secretory cells manufacture essential oils according to the plant's needs. Predation is likely to increase production of repellent essential oils, which may partly account for the extra deliciousness of organically grown herbs. If essential oil is produced as an attractant for pollinators, it may only be produced at certain times of day or night when the pollinators are most likely to be about.

Plant physiology

Let us now return to the molecular level and see how plant molecules are made. Plant tissues, like those in all living things, are made up of cells. Cells are made up of thousands of different types of molecules, each with a specific function in the cell's life. Millions of chemical reactions take place every minute inside each cell and most molecules in living cells exist for only a very short period before they are changed by some reaction or rearrangement of structure. You could say that life is the sum of all the chemical reactions happening in the cells of an organism at any one moment.

Plant cells have many substructures or organelles, for example a nucleus, mitochondria and chloroplasts. Each organelle has a specific function. The nucleus contains the genetic instructions for making proteins. Mitochondria are responsible for energy production. Chloroplasts containing the green pigment chlorophyll harness energy from sunlight for photosynthesis.

Photosynthesis

Photosynthesis is an important chemical reaction that takes place in plants and other organisms that don't survive by eating other creatures. It is the means by which plants obtain the carbon atoms that they need to create their molecules. Carbon dioxide from the air provides the carbon atoms, which join together with hydrogen and oxygen atoms (from water molecules) to make molecules known as sugars. Sunlight provides the energy needed for the reaction, which takes place in the chloroplasts found in leaves and green stems. Plants also need nitrogen atoms to make proteins and nucleic acids. Nitrogen from the air is mainly trapped or 'fixed' by bacteria in the soil. It can then be absorbed by plant roots in the form of water-soluble nitrogen compounds.

The chemical reaction for photosynthesis can be written as follows, showing glucose as the final product. Light and chlorophyll are required before the reaction can proceed:

$$6CO_2 + 6H_2O \xrightarrow{\text{Light and chlorophyll}} C_6H_{12}O_6 + 6O_2$$

Glucose might be the end product of photosynthesis, but the reaction initially forms more simple sugars. These simple sugars are then changed into many different types of molecules, including glucose. Glucose is a 6-carbon sugar used by every plant cell as an energy source. Each time a glucose molecule is broken down it releases the chemical energy it derived from sunlight during photosynthesis.

Secondary products

Plants can manufacture or absorb all the molecules they need for cellular functioning but they still have to solve another problem: how to defend themselves against predators. Some make defensive structures, such as thorns or spikes, but a large number prefer to use chemical 'weapons'. The defensive chemicals are known as 'secondary products', as they are not necessary for primary cellular functions, and essential oils form part of this chemical arsenal. Examples of secondary plant chemicals found in essential oils are:

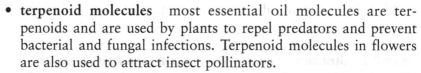

- **terpenoid molecules** most essential oil molecules are terpenoids and are used by plants to repel predators and prevent bacterial and fungal infections. Terpenoid molecules in flowers are also used to attract insect pollinators.
- **phenolic and phenyl propanoid molecules** a few phenolic compounds are found in essential oils. They are most likely used by the plant to repel predators.
- **non-terpenoid aliphatic molecules** these are also likely to repel predators, as they are found in citrus peel oils.
- **heterocyclic molecules** these molecules contain atoms other than carbon in closed rings. They are found in only a few oils, and include molecules such as the nitrogen-containing indole and methyl anthranilate, and the oxygen-containing lactones, coumarins and furanoid compounds.

Other types of secondary compounds found in plants include alkaloids, flavonoids and saponins. Several of these have notable pharmacological effects on humans, for example, the well-known alkaloid caffeine found in tea and coffee. However, such compounds are not extracted into essential oils by steam distillation.

How plants make essential oil molecules

Terpenoid molecules

Terpenoid molecules are made via the *mevalonic acid* biosynthetic pathway. This is the same pathway for the formation of carotenoid and sterol molecules (30 or more carbon atoms) and is also the pathway in animals for the formation of cholesterol and steroid hormones. The first step is to make mevalonic acid molecules (with six carbon atoms), hence the name of the pathway. Mevalonic acid molecules are then rearranged by a series of enzymatic transformations to form molecules known as isopentenyl pyrophosphate. Isopentenyl pyrophosphate consists of a branched 5-carbon molecule (known as an *isoprene unit*) joined to two phosphate groups. The isoprene unit is the starting point for manufacture of terpenoid compounds, the main type of molecule found in essential oils. Figure 2.3 shows the structure of an isoprene molecule and an isoprene unit. The illustration of the unit demonstrates the branching structure of the 5-carbon atom chain, but does not specify double bonds. The double bonds rearrange during the formation of terpenoid molecules, but isoprene molecules usually contain at least one C=C bond.

Figure 2.3   Structure of an isoprene molecule and an isoprene unit

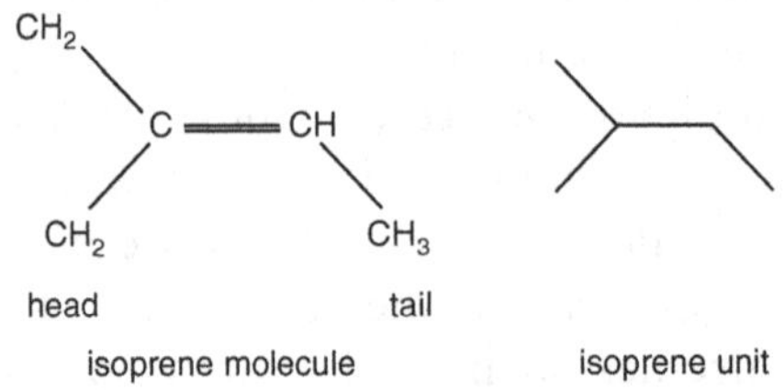

head   tail
isoprene molecule   isoprene unit

Isoprene molecules are easily linked together to form long, branched chains of carbon atoms. When rubber was first analysed in the late 1800s, it was found to contain hundreds of isoprene units all linked together. Rubber molecules are called *polymer* molecules (from the Greek *poly*, 'many' and *mer*, 'unit').

Terpenoid molecules in essential oils have either two or three isoprene units. The name 'terpene' derives from 'turpentine', a liquid solvent derived from the resin of Pinaceae species. When the structure of terpenoid molecules was first discovered, researchers thought the simplest molecule was one containing ten carbon atoms and started the naming system at this point, calling these molecules *monoterpenes* (from the Latin *mono*, 'one'). The later discovery of isoprene units (an isoprene molecule has five carbon atoms) has not changed use of the earlier naming system, so 'monoterpene' still means two isoprene units joined together. Terpenoid molecules with three isoprene units are known as *sesquiterpenes* (from the Latin *sesqui*, 'one and a half'). There are sometimes trace amounts of *diterpenes* in essential oils with four isoprene units, but the *triterpenes* with six isoprene units are waxes and are not found in steam-distilled essential oils because they are not volatile enough.

Monoterpenes

Monoterpenes contain two isoprene units, that is 10 carbon atoms. Figure 2.4 shows an example of myrcene, a common monoterpene in essential oils. Myrcene molecules follow the *isoprene rule* which says that isoprene units are usually joined 'head to tail' to make up the carbon skeletons of terpenoid molecules (Morrison & Boyd, 1987).[1] The number of C=C double bonds in terpenes varies from the exact number in each isoprene molecule as rearrangements occur during formation of the bonds between the isoprene units.

Figure 2.4 The monoterpene alpha-myrcene. The dotted line represents the bond formed between two isoprene units.

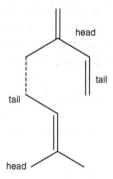

Sesquiterpenes

Sesquiterpenes contain three isoprene units, that is, 15 carbon atoms. Figure 2.5 shows the sesquiterpene (E)-beta-farnesene, an open chain sesquiterpene. However, one of the features of sesquiterpenes is that they readily undergo reactions that create molecular structures with closed rings. The molecular shapes formed by the closed rings can be classified into sub-groups of sesquiterpenes, for example *eudesmanes* and *azulenes*, but it is beyond the scope of this book to investigate these sub-groups beyond the detail found in Chapter 3.

Figure 2.5 The sesquiterpene (E)-beta-farnesene. The dotted lines represent the bonds formed between three isoprene units.

Both monoterpenes and sesquiterpenes can be modified by the addition of functional groups (see Table 1.4 on page 18 for a summary, and Chapter 4 for full descriptions). When a terpene has a functional group added to it, it is known as a terpen**oid**, either monoterpen**oid** or sesquiterpen**oid**. The **oid** ending means 'like', or 'derived from', hence 'terpenoid' can refer to all molecules with a terpene-like structure.

Some authors refer to all terpenoid molecules by the general term 'terpenes'. In this book, 'terpenes' are mono- and sesquiterpene hydrocarbons, molecules with only hydrogen and carbon atoms. Some citrus essential oils which have had their monoterpenes extracted by fractional distillation are sold as 'terpeneless' oils. Terpeneless oils are made up mainly of oxygenated terpenoid, aliphatic and coumarin molecules, and generally have a much stronger odour than the original essential oil. Not much research has been carried out yet on the physiological effects of terpeneless oils.

Phenols and phenyl propanoids

Phenols and phenyl propanoids are made via the *shikimic acid* pathway in plants which also make compounds like the tannins found in tea. As we have seen in Chapter 1, they have a distinctive benzene or aromatic ring. Phenols have a hydroxyl (–OH) group attached to the ring, and usually an isopropyl tail (3-carbon chain bonded to the ring at the middle carbon atom). Phenyl propanoids usually have a methyl ether functional group attached to the ring (see 'Phenyl methyl ethers' in Chapter 4), and a propenyl tail (3-carbon chain with one C=C bonded to the ring by one end). Figure 2.6 illustrates these differences.

Figure 2.6   Structure of phenols and phenyl propanoids. Thymol is a phenol, anethole is a phenyl propanoid.

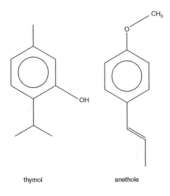

thymol   anethole

Non-terpenoid aliphatic molecules

The word 'aliphatic' describes molecules made up of carbon chains that are in a straight line and do not have a closed or aromatic ring. Examples of aliphatic molecules are the acrid-smelling C8 (8-carbon), C9 and C10 aldehydes found in small amounts in citrus oils, and the green-leafy smelling C6 compounds found in some floral oils like Jasmine and Rose. Figure 2.7 shows a molecule of the aldehyde octanal ($C_8H_{16}O$), found in Sweet orange oil. Aliphatic molecules are usually found only in trace amounts in essential oils, but if they have oxygenated functional groups attached their odours are usually noticeable despite this.

Figure 2.7   Structure of the aliphatic aldehyde octanal ($C_8H_{16}O$)

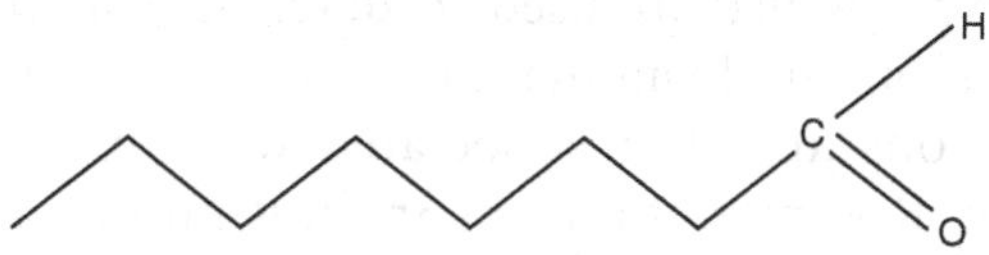

Heterocyclic compounds

Heterocyclic compounds are made up of carbon atoms arranged in a ring, with either a nitrogen or oxygen atom included as part of the ring. These molecules are uncommon in essential oils, occurring mainly in heady floral oils like Jasmine, Neroli and Narcissus. Figure 2.8 shows the structure of indole, found in Jasmine oil. Alkaloids are heterocyclic compounds that feature a nitrogen atom as part of the closed ring. However, alkaloid molecules are seldom found in steam-distilled essential oils as they are soluble in water.

Figure 2.8   Structure of the heterocyclic molecule indole, showing the position of the nitrogen atom

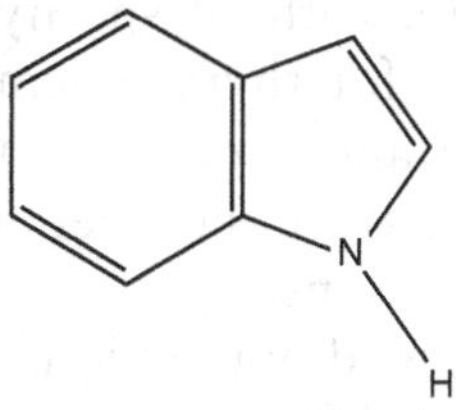

Variation in essential oil composition

While there are obvious variations in essential oil composition between oils from different species of the same genus (e.g. Peppermint and Spearmint), there are many factors that influence the composition of oils from different specimens of the same species. The most influential is that of geo-climatic location, as it gives rise to different chemotypes of essential oils. Other factors that affect the ratios of essential oil molecules made by the plant include soil type, life-stage of the plant (pre- or post-flowering) and even the time of day when harvesting is done.

Chemotype and effects of geo-climatic location

The term 'chemotype' is usually used to describe essential oils that vary in composition but are from the same species of plant. Different geographical locations are often associated with different chemotypes. For example, Rosemary oil from Spain is known as CT1 (CT, 'chemotype'), and has higher levels of camphor, whereas Rosemary oil from Tunisia is known as CT2 and contains higher levels of 1,8-cineole. A third rosemary chemotype, CT3 from France, has higher levels of verbenone, which is thought to be a less toxic ketone than camphor. Not surprisingly, due to the similarity of its climate with Tunisia, Moroccan Rosemary oil is usually the CT2 chemotype. Table 2.2 shows the varying ratios of major constituents in rosemary CT1 and rosemary CT2. It is thought that CT1's higher percentage of camphor and alpha-pinene makes it more effective for muscular aches, whereas CT2's higher percentage of 1,8-cineole is better for respiratory ailments.

Thyme (*Thymus vulgaris*) plants also produce at least five different chemotypes, although in this case chemotypical variation is not as related to geographical source.

The importance of knowing which chemotype you are using is even greater with thyme oils. Most aromatherapy books caution the use of Thyme oil on the skin, without specifying which chemotype. While this is a valid caution for thyme chemotypes with high percentages of phenols (thymol and carvacrol), the thyme chemotypes containing linalool and geraniol can be safely used on the skin at the usual aromatherapy concentrations.

Thyme oil harvested from wild populations (*Thymus vulgaris* Population) is often a mixture of different chemotypes, sometimes even

Table 2.2 Variation in the composition of Rosemary essential oil, chemotypes 1 and 2

	Rosemary CT1 (Spain)[1] %	Rosemary CT2 (Tunisia)[2] %
Alpha-pinene	22	10
Beta-caryophyllene	2.5	1
Beta-pinene	5	4
Borneol	2	8
Bornyl acetate	1.5	0.8
Camphene	9	3
Camphor	17	11
1,8-cineole	17	51
Limonene	4	2
Myrcene	4	1
Terpinen-4-ol	1.5	1
Verbenone	4	0.05

[1] M.H. Boelens (1985), 'The Essential Oils from *Rosmarinus officinalis* L.', *Perfumer & Flavorist* Oct./Nov. 10, pp. 21–37.

[2] G. Fournier, J. Habib, A. Reguigui, F. Safta, S. Guetari and R. Chemli (1989), 'Etude de divers echantillons d'huile essentielle de Romarin de Tunisie', *Plantes Medicinales et Phytotherapie* 23, pp. 180–5.

of different species. The odour of each batch should be carefully assessed to gauge the proportions of phenols it contains. Table 2.3 illustrates the differences in odour and therapeutic properties between the different chemotypes.

Another example of chemical variation is between Geranium (*Pelargonium graveolens* L'Herit. ex Ait) essential oil produced in the Reunion Islands (known as 'Bourbon' geranium) and Chinese geranium oil (see Table 2.4). Other countries also produce geranium oils with varying levels of constituents, but so far there have been no official chemotype numbers assigned as for the rosemary oils. This is perhaps because the oils from different areas consistently produce oils of the same chemotypic profile.

Other plant species that show chemotypical variation include:

Table 2.3   Comparison of four *Thymus vulgaris* chemotypes[*]

Chemotype	Major constituents	Description	Properties
CT thymol	Thymol para-cymene	Pungent, camphoraceous, herbaceous, clear to yellowish in colour.	Strong anti-bacterial, possible irritant.
CT carvacrol	Carvacrol Thymol borneol	Pungent, camphoraceous, red in colour (due to the carvacrol).	Strong anti-bacterial, possible irritant.
CT linalool	Linalool terpinen-4-ol linalyl acetate	Similar odour to Lavender oil (*Lavandula angustifolia*), clear to yellowish colour.	Mild anti-bacterial, possible sedative due to linalool content, non-irritant.
CT geraniol	Geraniol geranyl acetate	Similar odour to Geranium oil (*Pelargonium graveolens*) but more herbaceous, clear to yellowish colour.	Mild anti-bacterial, non-irritant.

[*] The list of constituents is taken from D. Pénoël and P. Franchomme (1990), *L'aromathérapie exactement*, Roger Jollois, Limoges, pp. 402–4.

basil, *Salvia* sp., *Cinnamomum camphora* and many *Eucalyptus* and *Melaleuca* species (see Brophy & Doran, 1996 and Webb, 2000 in 'Further reading' for more details on Australian essential oil chemotypes).

Simon et al. (1984) studied variations in essential oil composition due to soil type and rain-fall. They noted that clary sage (*Salvia sclarea*) plants require a dry chalky soil for maximum essential oil production, yielding very little oil if the soil is too rich. Lemon balm (*Melissa officinalis*) plants require deep soil and just the right amount of water. Too much or too little water resulted in reduced oil quantity and quality.[2]

Harvesting time

As all farmers know, there is an optimal time for harvesting crops. The same goes for essential oil producing plants. In a study done by Basker and Putievsky (1978) plants from the Lamiacae family

Table 2.4 Comparison of Geranium oils from China and Reunion[*]	Geranium, China %	Geranium, Reunion (Bourbon) %
Cis-rose oxide	1.9	0.6
Citronellol	40	22
Citronellyl formate	11	8
Geraniol	6	17
Geranyl formate	2	7.5
Iso-menthone	6	7
Linalool	4	13
Menthone	1.4	1.5
Trans-rose oxide	0.6	0.2

[*] G. Vernin, J. Metzger, D. Fraisse and C. Sharf (1983), 'Étude des huiles essentielles CG-SM Banque SPECMA: Essences de Geranium (Bourbon)', *Parfum. Cosmet. Arom.*, 52, pp. 51–61.

were tested to discover the optimum time for harvesting.[3] Their conclusions were that:

- the volatile oil content of the leaves increases with time, and also with the size of the leaf;
- maximum leaf yield was late summer for most Lamiaceae herb species, but the season for maximum oil yield and composition varied from species to species.

For example Sage oil (*Salvia officinalis*) contains different amounts of the neurotoxic ketone, alpha-thujone, depending on when it is harvested. Alpha-thujone taken internally can have neurotoxic effects on the central nervous system (GABA-antagonism), causing hallucinations in low doses and convulsions in larger doses (Tisserand & Balacs, 1995).[4] Sage leaves contain more alpha-thujone after the plant has flowered, so sage is usually harvested before flowering.

Flower oils like Jasmine are also affected by choice of harvesting time. Ahmad et al. (1998) examined the ratio and type of constituents produced by jasmine over a 24-hour cycle. Oil from flowers harvested in the morning contained the preferred combination of linalool, benzyl alcohol, cis-jasmone and indole, whereas those harvested in the evening contained higher levels of eugenol, benzyl benzoate and methyl salicylate. Eugenol and methyl salicylate introduce unpleasant odour notes which would not be acceptable in either the perfumery or aromatherapy industries.[5]

A brief overview of the essential oil industry

Over 3000 plant species produce essential oils, but only about 300 of these are available commercially. According to RIRDC (Rural Industries Research and Development Council, Australia), world trade in essential oils is growing hugely. In 1986, world imports and exports of essential oils and related perfumes and flavours were valued at a total of US$4157 million. In 1998 it was US$14 246 million. Australia currently accounts for only 1 to 2 per cent of the world trade in essential oils. The combined value of Australian imports and exports of essential oils has increased from US$34.3 million in 1986 to US$88.1 million in 1998.[6] According to B.J. Lawrence, the industry values of the top three essential oils in 1993 were: Sweet orange (US$58.5 million), Cornmint (*Mentha arvensis*) (US$34.4 million) and Eucalyptus (cineole type) (US$29.8 million).[7]

While the flavour and fragrance industry largely dominates the demand for essential oils (and hence controls their availability), novel oils produced on a small scale can still find their niche in the aromatherapy market. As we saw in Table 2.3, a knowledge of essential oil chemistry can equip aromatherapists to experiment safely with new oils and maybe even to estimate their possible therapeutic actions.

Further reading

- For more details on plant classification systems see: A. Cronquist (1988), *The Evolution and Classification of Flowering Plants*, New York Botanical Garden, Bronx, New York; and A. Takhtajan (1997), *Diversity and Classification of Flowering Plants*, Columbia University Press, New York.
- Photosynthesis—an in-depth introduction to university-level photosynthesis. http://photoscience.la.asu.edu/photosyn/education/photointro.html.
- Plant secondary metabolites—*Functions of Plant Secondary Metabolites*

and Their Exploitation in Biotechnology (1999), ed. Michael Wink, vols 2 and 3, Culinary and Hospitality Industry Publications Services, Weimar, Texas.

- Details on therapeutic properties of alkaloids and other plant products used in herbal medicine are well explained in S. Mills and K. Bone (2000), *Principles and Practice of Phytotherapy*, Churchill Livingstone, Edinburgh.

- A succinct university-level description of the formation of terpenes from isoprene units showing all the molecular details can be found at: http://www.usm.maine.edu/~newton/Chy251_253/Lectures/ BiopolymersI/TerpenesFS.html [accessed 25 May 2003].

- For useful information on variations in plant growth that happen under different cultivation methods, see *Plants in Action: Adaptation in Nature, Performance in Cultivation* (1999), ed. Brian Atwell, Macmillan, Australia.

- For some lovely microscopic images of essential oil secretory structures in plants see K.P. Svoboda, T.G. Svoboda and A.D. Syred (2000), *Secretory Structures of Aromatic and Medicinal Plants. A review and atlas of micrographs*, Microscopix Publications Knighton, Powys, UK.

- For details of chemical composition and chemotypic variation in Australian plants see J.J. Brophy and J.C. Doran (1996), *Essential Oils of Tropical Asteromyrtus, Callistemon and Melaleuca species*, ACIAR Monograph No. 40, Australian Centre for International Agricultural Research, Canberra; and M.A. Webb (2000), *Bush Sense: Australian Essential Oils and Aromatic Compounds*, self-published, Australia, ISBN 0646 40353 2.

- For further details of global essential oil production see B.M. Lawrence (1993), 'A planning scheme to evaluate new aromatic plants for the flavor and fragrance industries', in J. Janick and J.E. Simon (eds), *New Crops*, Wiley, New York, pp. 620–7, http://www.hort.purdue.edu/ newcrop/proceedings1993/v2-620.html [accessed 25 May 2003].

- For information on essential oil production in Australia see ESSENTIAL OILS & PLANT EXTRACTS Research Program, Rural Industries Research & Development Corporation (RIRDC), Canberra, Australia, http://www.rirdc.gov.au/programs/eop.html [accessed 25 May 2003]. The New Crops newsletter can be found at http://www.newcrops. uq.edu.au/newslett/ncnl1218i.htm.

TERPENES

This chapter looks at terpenes, which form the largest class of molecules found in essential oils. Terpenoid chemistry is a branch of organic chemistry. As we have seen, monoterpene and sesquiterpene molecules are made up of hydrogen and carbon atoms, with carbon atoms forming the skeleton or structural shape of the molecule. When an oxygen atom or oxygen-containing functional group is added to either of these types of terpene molecule, we call it a terpenoid molecule.

Our goal is to come to an understanding of what sort of interactions terpenoid molecules might have with molecules in the human body. The concepts of polarity, solubility and emulsification are useful when considering how terpenoids might react once inside the body.

Physical characteristics of terpenoid molecules

To start with, let's look at what sort of substances terpenoid molecules are. On a physical level, they have the following characteristics (which are shared by non-terpenoid compounds in essential oils like phenylpropanoids):

- Volatility—They evaporate easily, mostly below 100°C. This means that when they are applied to the skin, they will to some extent evaporate with the warmth of the skin unless the site of application is covered. It also means that inhalation is an effective application method.
- Flammability—Terpenoids in bulk are labelled 'flammable liquids', though they are not nearly as flammable as other organic fuels such as octane.
- Density—Mono- and sesquiterpenoids are nearly all less dense than water, meaning they will float on the surface of water.

- Odour—Most terpenoid molecules in essential oils can be perceived by the olfactory system. Some are pleasant, others unpleasant. There is a great range in thresholds of odour perception for different molecules. Some substances, like rose oxide (found in *Rosa damascena*), may be detected at a concentration of only 0.5 parts per billion (ppb). Others, like nerol, also found in *Rosa damascena*, can be detected at 300 ppb (Dodd, 1988).[1]

It is of interest to aromatherapists how terpenoid molecules behave when they come in contact with the substances found in human bodies, such as skin oils, cell membranes and the variety of substances in blood and other body fluids. The types of environment found in the body can be identified as either hydrophilic (water-loving) or lipophilic (fat-loving). The blood, inter- and intra-cellular fluids and urine are all mainly composed of water and therefore hydrophilic. Skin oils and cell membranes are made of lipids and are lipophilic. To understand how essential oils might interact with hydrophilic or lipophilic molecules, let's examine the concepts of polarity, solubility and emulsification.

Polarity

Polar molecules

If we go back to Chapter 1, where we were looking at molecular structure, we notice that molecules are usually neutrally charged. This is because the numbers of protons and electrons are equal in each component atom, and there is no imbalance of charge. However, molecules of some substances seem to behave as though they are actually charged, and undergo attraction and repulsion with each other.

Water is one of these substances. Instruments which can measure the overall charge over a molecule of water detect that near the oxygen atom there is an area of slight negative charge, and near the two hydrogen atoms there are areas of slight positive charge. Figure 3.1 shows a water molecule with its different areas of charge.

The imbalance in charge is ascribed to the presence of the oxygen atom. Due to the size of oxygen atoms and the way their electrons are arranged, oxygen atoms are strongly '*electronegative*'. Electronegativity is the relative ability of an atom to attract electrons towards its nucleus. So when oxygen atoms form bonds with other atoms with low electronegativity, like hydrogen, the electrons in the

Figure 3.1   Polarisation of charge in a water molecule

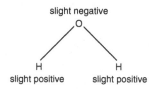

bonds are drawn more closely towards the oxygen atom, rather than being equally shared between the two atoms. Other types of highly electronegative atoms are fluorine, nitrogen and chlorine. Both carbon and hydrogen have low electronegativity.

Most molecules containing electronegative atoms are *polar* molecules. The electronegative atom creates an area of relative negative charge, or negative pole, and the other atoms attached, being relatively depleted of electrons, function as positive poles. These poles behave like the north and south poles on a magnet. For interested readers, any university-level organic chemistry text should be able to expand the topic of polarity further.

When water molecules are together as a liquid, the positive and negative poles are attracted to each other, forming weak liaisons known as *electrostatic liaisons* or *hydrogen bonds*. These are not true bonds, because there is no sharing of electrons.

The term 'hydrogen bonding' comes from the fact that most often it is a hydrogen atom attached to an electronegative atom that becomes the positive pole. Because these hydrogen atoms are at the surface of molecules, they tend to act like 'sticky' ends and seek out electrostatic connection with electronegative atoms on other molecules, or with ions in solution. As hydrogen atoms are nearly always at one end of the liaison, the liaisons have become known as hydrogen bonding. In Figure 3.2 the electrostatic liaisons between the water molecules are represented by broken lines. The slight negative (δ-) and slight positive (δ+) charges act like magnets.

The hydrogen bonding of water forms a kind of lattice. A way of imagining hydrogen bonding between water molecules is to picture the water molecules as members of a secret society refusing admittance to anyone who cannot do the secret handshake, that is, participate in the attractions and repulsions made possible by the polarity of the molecules. Electrostatic liaisons also allow water molecules to exist in a lower energy state, increasing their stability.

Figure 3.2   Three water molecules with electrostatic liaisons represented by broken lines

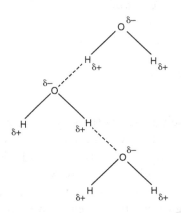

Water is the only completely polar molecule we will deal with in this book. Essential oil molecules with oxygen-containing functional groups, like alcohols, phenols, aldehydes and ketones are also polar at the site of the oxygen atom. The presence of an exposed electronegative oxygen atom increases a molecule's interactive power for therapeutic or hazardous properties, as we will see in Chapter 4.

Non-polar molecules

Non-polar molecules are molecules in which there are no strongly electronegative atoms, or if there are, their electron-drawing powers cancel each other out. Terpenes are made up of carbon and hydrogen atoms, neither of which are strongly electronegative. This means that the charge over the molecule is neutral, and there is no development of negative or positive poles. The molecules are therefore called *non-polar*. Hydrocarbon molecules such as vegetable oils and fats are examples of non-polar molecules.

Although non-polar substances cannot form electrostatic lattices like water, there is a weak force which keeps the non-polar molecules of a substance connected to each other. This force is made up of momentary attractions between the molecules. These momentary attractions are caused by the continual movement of the electrons in the molecular energy path (the composite energy path derived from bonds between the atoms in the molecule). They are very weak and transitory attractions, and only act when the molecules are in liquid form.

Solubility

The property of solubility is one of the most important for aroma-therapists to understand, as it applies not only to the use of different solvents in blending, but also to the mechanism of epidermal penetration of the essential oils, and eventual transport of essential oils in the blood. For a substance to be dissolved in another substance (the solvent), the molecules of each substance must freely co-mingle. As a general rule of thumb, polar substances will dissolve in polar solvents, and non-polar substances in non-polar solvents.

Solubility of terpenoids in vegetable oils

When non-polar terpenoids are mixed with non-polar vegetable oils they disperse completely in each other. This is why essential oils dissolve in substances like vegetable oils, mineral oils and, to some extent, full cream milk. Substances which dissolve in fats and oils are called *lipophilic*, literally meaning they 'like fats' (from Greek *lipos*, 'fat'; *philos*, 'loving').

Solubility of terpenoids in water

The question 'Why don't essential oils and water mix?' can be answered fairly simply if you bear in mind that all molecules prefer to stay in the lowest energy state possible. As a general rule, the more complex a structure, the more energy is required to hold it at that level of complexity. So, when faced with a mixture of a polar substance such as water, and a non-polar substance such as an essential oil molecule, the following situations occur:

- Non-polar essential oil molecules cannot form electrostatic liaisons with water molecules. The presence of a non-polar molecule disrupts the low energy lattice of the water molecule electrostatic liaisons.
- The water molecules will therefore repel the terpenoid molecules, rather than let them enter the lattice.
- Being less dense than water, the terpenoid molecules float on the top of the water. Some non-terpenoid molecules, like the phenyl propanoid, eugenol, are more dense than water at normal room temperature. A drop of eugenol will sink to the bottom in a spherical droplet.

Monoterpenes and sesquiterpenes are non-polar hydrocarbons and therefore not soluble in water. However, several terpenoid molecules have oxygen-containing functional groups. If the oxygen atom is part of a hydroxyl functional group (–OH group), this part of the molecule becomes polar and can form liaisons with water molecules.

The difference that the addition of an –OH group makes to a hydrocarbon in terms of its solubility in water can be seen in the different solubility of the four smallest hydrocarbons—methane, ethane, propane and butane—and their corresponding alcohols (see Table 3.1). Also note that at room temperature, all the hydrocarbons are gases, whereas all the alcohols are liquids. The –OH groups help the molecules link together more strongly than is possible when they are only hydrocarbons.

Table 3.1 Comparison of solubilities of hydrocarbons and alcohols in water			
Alkane	Solubility, g/100g H_2O	Alcohol	Solubility, g/100g H_2O
Methane CH_4	0	Methanol CH_3OH	Infinite
Ethane C_2H_6	0	Ethanol C_2H_5OH	Infinite
Propane C_3H_8	0	Propanol C_3H_7OH	Infinite
Butane C_4H_{10}	0	Butanol C_4H_9OH	7.9 g/100g H_2O

Source: adapted from R.T. Morrison and N. Boyd (1987), *Organic Chemistry* 5th edn, Allyn & Bacon Inc., Boston, p. 227.

For a terpenoid molecule with 10 carbon atoms, such as the monoterpenoid alcohol, menthol, a single –OH group is not sufficient to allow the molecule to dissolve in water. However, molecules like glucose ($C_6H_{12}O_6$), which have several –OH groups, can quite readily dissolve in water. The –OH groups change the polarity of the molecule and act like a 'passport' into the polar lattice of the water. Figure 3.3 shows the structures of menthol and glucose. The darker lines in the glucose structure indicate the three-dimensional structure of the ring. They are 'closer' than the thinner lines. Glucose is *hydrophilic* (Greek *hydro*, 'water', *philic*, 'liking'), whereas the menthol is *hydrophobic* (literally 'fears water'; Greek *phobic*, 'fearing').

Essential oil constituents with polar functional groups partially dissolve in water, and thus the boundary between the oil and water layers is less distinct. It is possible to achieve a temporary emulsion of water and some monoterpenoids with –OH functional groups by

Figure 3.3 Structures of menthol and glucose, showing water-soluble –OH groups

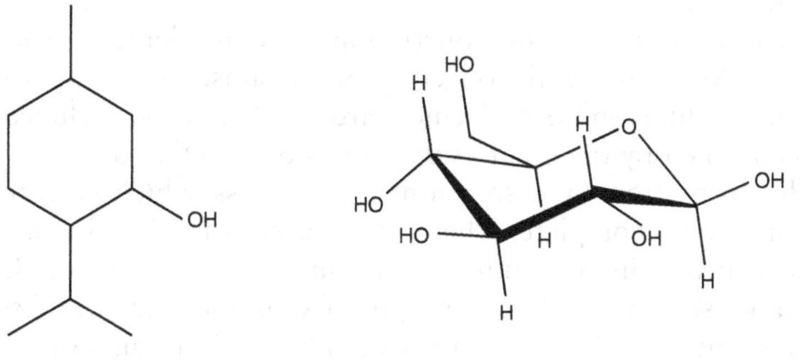

menthol

glucose

using hot water and shaking vigorously. Try Geranium oil, which contains a large percentage of the monoterpenol, geraniol. After shaking, the water is cloudy and the oil globules are no longer evident on the surface. In general, the longer the carbon chain, and the bulkier the terpenoid molecule, the less likely it is to be soluble in water. Sesquiterpenoids, with their lipophilic/hydrophobic 15 carbon atoms, are less likely to dissolve in water than monoterpenoids.

Solubility of terpenoids in ethanol

Ethanol (the alcohol found in beverages like wines and spirits) is a solvent often used to dissolve essential oils in perfumery. It has two carbon atoms and a hydroxyl group (see Figure 3.4). Ethanol is both a polar and a non-polar solvent, due to its –OH group on the one hand and its C_2H_5– group on the other hand. The –OH groups of the ethanol form a similar lattice to that of water, but it is not as tightly linked together.

Perfumers use solvents like ethanol because they dissolve essential oils and also dissolve in water, thereby allowing the non-polar

Figure 3.4 Structure of ethanol

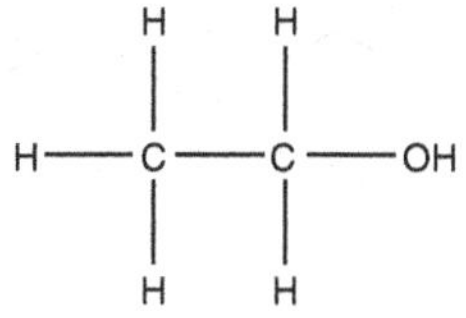

essential oil molecules to be incorporated into the water. This mechanism is known as *emulsification* and is discussed in detail below.

Most monoterpenoid constituents are moderately soluble in ethanol. Molecules with longer carbon chains, such as in sesquiterpenoid or diterpenoid molecules, are too big to be included in the lattice formed by the polar –OH groups of the ethanol.

The formation of a solution is a process which can be either complete or incomplete. There are degrees of solubility which are usually given in a volume-to-volume ratio (for example, 1:15 substance:solvent), or by the weight of substance that will dissolve in a given amount of solvent, for example, 2 g/L (grams of substance per litre of solvent). If you try and dissolve more of the substance in the solvent than the ratio indicates, it will not dissolve. In perfumery, a description of the degree of solubility is given, for example, 'soluble to a turbid or opalescent state'. Turbidity (or milkiness) is usually caused by the perfumery ingredient (or essential oil) forming an emulsion of very small opaque droplets in the water, giving it a cloudy appearance, as mentioned in the example with Geranium oil above.

Emulsification

An emulsifier is a molecule which dissolves in both water and oil. In order to do this, it has to have a non-polar lipophilic end, usually a carbon chain, and a polar hydrophilic end with one or more water-soluble groups attached. Emulsifiers allow normally insoluble lipophilic substances such as terpenoids and vegetable oils to mix with water, forming a lotion or cream, depending on the proportions of lipophilic substance and water. If it is an oil-in-water (o/w) emulsion (less oil, more water), the result will be a lotion. A water-in-oil (w/o) emulsion is more likely to be a cream.

Examples of emulsifiers include lecithin and modified vegetable

Figure 3.5 Schematic representation of an emulsifier molecule

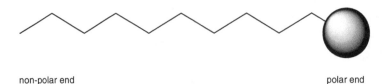

non-polar end   polar end oils which have several polar groups attached to one end of the molecule. Figure 3.5 is a schematic diagram of an emulsifier molecule, with the long zigzag representing bonds between carbon atoms and the circle representing a polar oxygen-containing functional group or cluster of –OH functional groups.

When emulsifiers are put into water, they turn the water milky as they are dispersed throughout the water without leaving an oily film on the surface. On the molecular level, the emulsifier molecules gather together and form tiny hollow spheres or *micelles* inside the water lattice, with their polar ends all facing outwards and their non-polar ends tucked safely inside the micelle. Figure 3.6 is a schematic representation of a micelle of emulsifier molecules, using the schematic of Figure 3.5 to represent the emulsifier.

Figure 3.6 Cross-section view of a micelle of emulsifier molecules: non-polar ends in centre of sphere, polar ends on surface of micelle able to liaise with polar water molecules

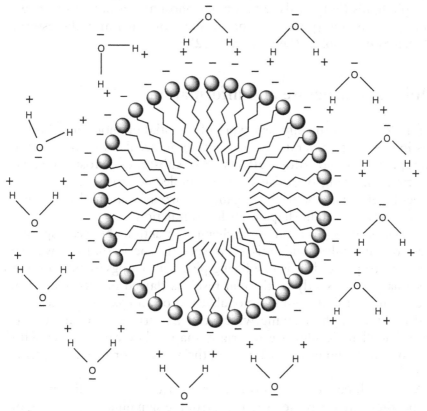

Terpenoid molecules and other non-polar molecules can be incorporated into the centre of micelles and thus 'solubilised' in water. Detergents are also emulsifiers, absorbing the grease from cooking implements or body oils from clothes and incorporating them into micelles of detergent molecules. In a similar way, emulsifying substances in the bile help emulsify dietary fats in the upper small intestine and aid their absorption into the bloodstream.

Polarity and solubility in the body

The purpose of discussing polarity and solubility is to give an understanding of how essential oils might behave in a largely watery environment such as the human body. Essential oils, being lipophilic, are likely to diffuse easily into and out of cells, as they can dissolve in the lipophilic cell membranes. However, no systematic investigation of essential solubility in body fluids has yet been conducted. It is likely that essential oil molecules are rapidly emulsified by lipoproteins in body fluids, as lipoproteins are complex micelles of phospholipids (fatty acids with a polar phosphate group at one end), globulin proteins (polar and non-polar portions) and cholesterol molecules (non-polar) (Lehninger, 1982).[2]

Molecular structure and information

Before looking at the different types of essential oil molecules, I think it is important to reiterate the influence of molecular structure on the therapeutic or hazardous effects of a substance. Most chemical reactions in living organisms require molecules to be in a certain three-dimensional shape. For example, enzymes are molecules that need to have a certain shape in order to function.

Enzymes work by facilitating chemical reactions between specific sorts of molecules, and it is their specific shape that allows the reactions to proceed. It is like fitting pieces into a jigsaw puzzle—the piece has to fit exactly or the end result is flawed or incorrect. Any alterations in the three-dimensional shape of an enzyme can lead to malfunction of the enzyme which can in turn lead to disease. If essential oil molecules are the right shape, they can interact with enzymes in the body, either blocking their action, or promoting their action.

Most cellular receptor proteins also are very specific in their shape; these are the proteins that control the signalling and passing of information between cells. An example of cells with receptor proteins are olfactory nerve cells. These cells can detect the presence of very specific shapes of odour molecules and then pass on information to the brain. The brain then interprets the information to mean that particular substances are present in the air. It is possible other receptors in the body respond as sensitively to differences in terpenoid structure as do olfactory receptors.

Different types of terpenoid molecules can be classified into groups by their carbon chain structure and the presence or absence of oxygen atoms and arrangements of bonds. In the remainder of this chapter, and Chapter 4, I examine the various structures of essential oil molecules, and evidence from the scientific literature about their therapeutic or hazardous effects. My theory is that the molecular structure of an essential oil constituent may indicate its therapeutic or hazardous effect on the body. And if you know the structure of the major constituent of an oil, you can hazard a guess as to the likely major effects of that oil.

However, most essential oils contain a number of different constituents that may have complex interactions once inside the body. These interactions have not yet been studied in any detail. It is possible that some constituents in an oil might have antagonistic properties, for example, some being irritating and others anti-inflammatory. I consider the information in Chapters 3 and 4 about the different chemical constituents to be like a series of clues in a detective novel. It requires some lateral thinking and earnest enquiry to make sense of the mystery, but in the end you may come up with some useful hypotheses about how essential oils work on the body.

Figure 3.7 is like a road map, showing how the different types of essential oil molecules are chemically related to or derived from one another. The solid lines are definite chemical pathways, whereas the broken lines involve complex rearrangements, and are only suggestions as to how the molecular structures are related. The molecular fragments shown are the functional groups (compare Table 1.3).

It is interesting to note that a very small change to a functional group, such as the alteration of a C–OH group to a C=O group, can cause a considerable difference in the odour and therapeutic properties of a molecule, while the carbon chain structure remains the same. In terms of the therapeutic power of the different groups, it seems that there is a general increase as you progress down the road map. In other words, the monoterpenes and sesquiterpenes are not as powerful, nor are their effects as profound, as the monoterpenols and sesquiterpenols. I consider the lactones and coumarins right at the bottom of the page to be the most powerful.

By 'powerful' I mean that even a small percentage of the substance in an oil has a noticeable effect on the overall properties of the oil. An example is the furocoumarin, bergaptene. Though present at less than 1 per cent in cold-pressed Bergamot oil, bergaptene is nevertheless powerful enough to cause a dermatitis-like reaction in sunlight when the oil is applied neat to the skin. The total bergaptene content of cosmetic products is recommended to be less than 0.0015 per cent.

Figure 3.7   Essential oil chemistry road map

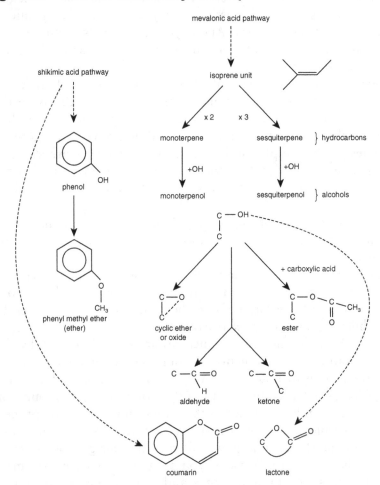

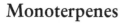

Monoterpenes

Structure

Monoterpenes are unsaturated hydrocarbons which have 10 carbon atoms (two isoprene units), with at least one double bond. A carbon chain molecule with no double bonds between carbon atoms is known as a *saturated* molecule, because all the carbon atoms are 'saturated' with hydrogen atoms (that is, they have only single bonds), and thus have little power to react without input of free energy. *Unsaturated* molecules have at least one C=C double bond.

It appears that molecules which have several double bonds are more chemically reactive than constituents with no double bonds. One of the features of double bonds is that they can undergo a reaction whereby one of the bonds is broken, and two other atoms or parts of molecules can be added to the two carbon atoms which were previously double bonded.

The number of hydrogen atoms in monoterpenes varies depending on the number of double bonds. The more double bonds, the fewer hydrogen atoms. Monoterpenes can have either open chain, mono-cyclic or bicyclic structures. The closed rings of the monocyclic and bicyclic structures give the molecules greater complexity, which may increase their specificity for different chemical reactions in the body.

Distribution of monoterpenes in common essential oils

The values for the distribution of different compounds in various essential oils, shown in the tables in this chapter and Chapter 4, are average percentage values, rounded to the first decimal point. The data in these tables are amalgamated in the Ready Reference List at the back of the book and share the same references unless otherwise noted. The percentages come from gas chromatographic and mass spectra analyses of single batches (see Chapter 6). Other aroma-therapy books give a range of percentages reflecting the acceptable variations in composition from several batches.

Naming

All monoterpenes have the ending *-ene*. If you are looking at the name of a constituent, for example myrc*ene*, the *-ene* ending indicates the presence of at least one double bond between carbon atoms, as noted in Chapter 1 (see Table 1.3).

Table 3.2   Examples of monoterpenes

Monoterpene	Molecular structure
myrcene $C_{10}H_{16}$ open chain Bay (West Indian) 13% Juniper (berry) 11%	
alpha-pinene $C_{10}H_{16}$ bicyclic Cypress 45%* Pine 44.1% Scotch pine 42% Kunzea 39% Frankincense (Somalia) 34.5% Juniper (berry) 33% Angelica (root) 25% Rosemary (Spain) 22% Immortelle 21.7% Eucalyptus 14.7% Myrtle 8.1%	
limonene $C_{10}H_{16}$ monocyclic Grapefruit 93% Sweet orange 89% Mandarin 71% Lemon 70% Lime (Persian) 58% Caraway 46% Bergamot 38.4% Black pepper 14.5% Elemi 54%	

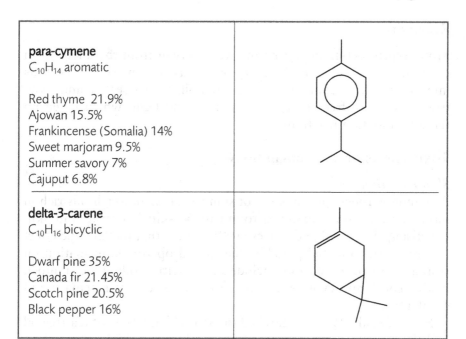

para-cymene $C_{10}H_{14}$ aromatic Red thyme 21.9% Ajowan 15.5% Frankincense (Somalia) 14% Sweet marjoram 9.5% Summer savory 7% Cajuput 6.8%	
delta-3-carene $C_{10}H_{16}$ bicyclic Dwarf pine 35% Canada fir 21.45% Scotch pine 20.5% Black pepper 16%	

* Pénoël D. and Franchomme, P. (1990), p. 345.

Solubility

Monoterpenes, as with most essential oil constituents, are not soluble in water, as they are hydrocarbons with long non-polar carbon chains. They are soluble in other oils, such as almond oil or other vegetable oils, which are also non-polar, and are fairly soluble in ethanol.

Volatility

Volatility refers to the ease with which a substance will evaporate. Smaller molecules with lower boiling points evaporate more rapidly than larger molecules with higher boiling points. If a constituent is highly volatile, it may evaporate before it has a chance to penetrate the skin. Monoterpenes are among the most volatile of the groups of constituents, and are often thought of as 'top' notes in perfumery, meaning that these are the odours that you first smell when you smell a blend.

Reactivity

Monoterpenes combine over time with oxygen from the air to form peroxides, epoxides and sticky resinous polymers. It is thus important to keep containers of essential oils tightly sealed, and away from sources of free energy such as heat and light, which speed up these degradation reactions.

Toxic effects on the human body

Skin irritation

Pure monoterpenes are mostly not skin irritants, although oils rich in monoterpenes are sometimes found to be skin irritating and skin sensitising. Tisserand and Balacs (1995) report that the 'allergic effect of terpene-rich oils [is] possibly due to hydroperoxide formation on storage'.[3] Terpene-rich oils include all the citrus oils, pine, juniper, black pepper and cypress oils, and to a lesser extent rosemary and eucalyptus.

Hausen et al. (1999) researched the sensitising effects of tea-tree oil and the monoterpenes it contains by patch-testing 15 human volunteers, and found that oxidised tea-tree oil, in particular the peroxides, epoxides and endoperoxides that were formed from the monoterpenes in the oil, were the sensitising agents.[4] Para-cymene (reported as a skin irritant by Tisserand & Balacs (1995, p. 189)) was also found to increase in the oxidised product, at the expense of alpha- and gamma-terpenines.[5]

Respiratory irritation

Alpha- and beta-pinene and delta-3-carene, all constituents of turpentine oil, have been investigated as likely respiratory irritants in sawmills in Sweden. The uptake of the constituents from the air was between 62–68 per cent, and after two hours' exposure, 2–5 per cent was excreted unchanged in expired air. The mean half-lives of the three constituents in the body were 32, 25 and 42 hours respectively for the test amount of 450 mg/m^3 of turpentine in the air.

At the end of the exposure, subjects experienced discomfort in the throat and airways, and airway resistance was increased after the end of the exposure (Filipsson, 1996).[6] Depending on the administration method, inhalations of pinene-rich oils may reach this concentration, so it is important to ask whether people suffer from asthma or other conditions which are already increasing their airway resistance before suggesting the use of such oils.

Possible kidney irritation

Pénoël and Franchomme (1990) suggested that turpentine and juniper twig essential oils give a 'definite nephrotoxicity'.[7] However, Schilcher and Leuschner (1997) tested rats which were fed up to 1000 mg of juniper oil for kidney damage, and found no functional or morphological changes.[8] Essential oils from some species of juniper (for example Savin, *Juniperus sabina*) contain the ester sabinyl acetate, which is embryotoxic in rats and may also be generally toxic (Tisserand & Balacs, 1995).[9] Juniper oil producers may confuse *Juniperus communis* and *Juniperus sabina*, but it is unlikely.

Therapeutic effects of monoterpenes

Mucolytic effect

Monoterpenes tend to have a 'drying' effect on skin and mucous membranes. This can feel like a tightening or 'tonic' sensation on the skin. The pinenes are noted by Pénoël and Franchomme (1990) as mucolytics (agents that cause mucus to disintegrate or dry up),[10] which is interesting in light of the airway irritation mentioned above. Pine oils have been used in respiratory infections presumably for this mucus-drying effect. However, I have not been able to find supportive evidence in the research literature for this assertion.

Gallstones

Limonene has been used as an agent to dissolve gallstones through direct injection into the biliary system. In a study by Igimi et al. (1991),[11] it was found to dissolve the gallstones completely in 48 per cent of cases and subsequent research indicates that it is most efficacious where the gallstones are cholesterol based (Igimi et al., 1992).[12] It is doubtful that the amounts used in a typical aromatherapy treatment would have any impact, but it is worth a try. Limonene is found in a great many essential oils, and in high percentages in the citrus peel oils.

Cancer prevention and treatment

This is an area receiving a great deal of research attention, and it has been found that several essential oil constituents show promise as cancer prevention agents. They work to prevent the initiation, promotion and progression of cancers, and also show promise as

Table 3.3　Essential oils containing high percentages of monoterpenes

Angelica root *Angelica archangelica*	alpha-pinene 25%	1,8-cineole 14.5%	alpha-phellandrene 13.5%
Elemi, *Canarium luzonicum*	limonene 54%	alpha-phellandrene 15.1%	elemol 15%
Grapefruit (Israel)	limonene 93%	myrcene 2%	alpha-pinene 0.6%
Lemon (Argentina)	limonene 70%	beta-pinene 11%	gamma-terpinene 8%
Lime (Persian)	limonene 58%	gamma-terpinene 16%	beta-pinene 6%
Mandarin (Italy)	limonene 71%	gamma-terpinene 18.5%	alpha-pinene 2.4%
Orange, sweet (Brazil)	limonene 89%	myrcene 1.7%	beta-bisabolene 1.3%
Fir balsam (Canada), *Abies balsamea*	beta-pinene 30%	delta-3-carene 21.5%	bornyl acetate 11.9%
Juniper berry	alpha-pinene 33%	myrcene 11%	beta-farnesene 10.5%
Nutmeg	alpha-pinene 22%	sabinene 18.6%	beta-pinene 15.6%
Frankincense (Somalia)	alpha-pinene 34.5%	alpha-phellandrene 14.6%	para-cymene 14%
Dwarf pine, *Pinus mugo* ssp. *pumilia*	delta-3-carene 35%	beta-phellandrene 15%	alpha-pinene 13.1%
Pine, *Pinus pinaster*	alpha-pinene 44.1%	beta-pinene 29.5%	myrcene 4.7%
Scotch pine, *Pinus sylvestris*	alpha-pinene 42%	delta-3-carene 20.5%	limonene 5.2%
Bergamot (Calabria)	limonene 38.4%	linalyl acetate 28%	linalool 8%
Rosemary (Spain) camphor chemotype	alpha-pinene 22%	camphor 17%	1,8-cineole 17%

therapeutic agents. Limonene in particular has been investigated, and the monoterpenol perillyl alcohol. See Chapter 7 for the structure of perillyl alcohol.

The cancers which have been treated in rats are breast and pancreatic carcinomas. Limonene and perillyl alcohol appear to inhibit the isoprenylation of small G-proteins, thus altering the regulation of certain genes, the products of which are responsible for blocking cancer cells' metabolic processes (Gould, 1997).[13] Clinical trials with humans are underway, but no conclusive results are available at the time of writing.

Muscular aches

Para-cymene, found in several oils, is specified for muscular aches by Pénoël & Franchomme (1990).[14] Its effectiveness is probably due to its mild irritant and rubefacient effect on the skin. Whenever I read of 'warming' or 'stimulating' oils, I would expect them to be possible irritants if applied neat to the skin. Rubefacient substances increase cutaneous bloodflow to the local area, which is helpful in the case of muscular and joint pain. Several of the other monoterpene constituents, particularly the bicyclic pinenes and delta-3-carene, are also noted as rubefacients or mild irritants.

Essential oils with high percentages of monoterpenes

Table 3.3, and similar tables in subsequent sections, is extracted from the table in the Appendix for easy reference. The constituents listed are those with the highest three percentages in each oil; constituents which are not monoterpenes are shaded in grey. Monoterpenes are present as major constituents in a variety of oils. The main odour characteristics are the woody-pine of the pinenes, the citrus of limonene and the medicinal-resin of the terpinenes.

Sesquiterpenes

Structure

Sesquiterpenes are hydrocarbon molecules which have 15 carbon atoms (three isoprene units), and a varying number of hydrogen atoms. There are no oxygen atoms.

Table 3.4 **Examples of sesquiterpenes**	
Sesquiterpene	**Molecular structure**
beta-caryophyllene $C_{15}H_{22}$ bicyclic Black pepper 34.6% Patchouli 20% Ylang ylang 10.5% Clove bud 9.8% Basil 7% Lavender (French) 5.16% Immortelle 5%	
chamazulene $C_{14}H_{16}$ bicyclic (not quite sesquiterpenoid; only 14 C atoms) German Chamomile 15%	
beta-farnesene $C_{15}H_{22}$ open chain German Chamomile 27% Juniper (berry) 10.5%	

Naming

Being terpenes, sesquiterpenes also end *-ene*, and, like the monoterpenes, the rest of the name is derived either from the type of plant it was found in, or from the country it was found in. For example, alpha-pinene was first found in pine oils (*Pinus* sp.), and chamazulene was first found in Chamomile oil. Some sesquiterpenes occur in only one or a few oils, for example patchoulene in Patchouli oil, cedrene in the Cedarwood oils and santalenes in Sandalwood oil.

A subset of sesquiterpenes known as *azulenes* have a dark blue or dark green colour. Both chamazulene from German chamomile oil, and guaiazulene from Blue Cypress (*Callitris intratropica*) and Yarrow oils, are present in sufficient quantities to give these oils their distinctive blue or green colour.

Solubility

Sesquiterpenes are not soluble in water, though they do dissolve readily in other oils and non-polar solvents. Due to their larger size they do not dissolve as readily in ethanol as monoterpenes.

Volatility

Due to their higher molecular weight and higher boiling points, sesquiterpenes are not as volatile as monoterpenes, and this also accounts for their presence as middle or base notes in perfumery. They are more viscous than the monoterpenes, and an oil high in sesquiterpenes will pour more slowly out of the bottle. Most of the woody oils tend to be sesquiterpenoid in character, containing high percentages of sesquiterpenes and sesquiterpenols. An exception is Rosewood (*Aniba rosaeodora*), which contains mainly linalool, a monoterpenol.

Reactivity

Similarly to monoterpenes, sesquiterpenes react over time with oxygen in the air, and if exposed to energy in the form of heat or light will form epoxides and alcohols and eventually polymerise to form long chain resins. In some oils, such as Patchouli and Vetiver, this oxidation is thought to improve the odour, and thus aged oils are often prized above newly distilled ones. I have not yet seen any research on comparison of the therapeutic effects of new and old oils.

Toxic effects on the body

I would expect sesquiterpenes to also become skin irritants when oxidised, but have not found any research reporting them as skin sensitisers or irritants. Keep oils with high percentages of sesquiterpenes out of contact with air and light as much as possible.

Therapeutic effects of sesquiterpenes

Anti-inflammatory effects

Sesquiterpenes with several double bonds are supposed to be good for reducing inflammation caused by stings and bites, and for histaminic reactions (Pénoël & Franchomme, 1990).[15] No research evidence has been found for this assertion.

Beta-caryophyllene has been found to reduce stomach cell damage from alcohol poisoning in rats when administered orally. It seems to work in a better way than standard non-steroidal anti-inflammatory preparations as it does not damage the gastric mucosa (Tambe et al., 1996).[16]

Chamazulene, found in German Chamomile oil, has been shown to be anti-inflammatory *in vivo*, and Safayhi et al. (1994) found by *in vitro* studies that the anti-inflammatory effect is due to the blocking of formation of leukotriene B4 in neutrophils.[17] Leukotriene B4, one of the inflammatory mediators released by neutrophils at the site of inflammation, in turn causes further inflammation. This may be a mechanism whereby chamazulene helps soothe insect bites, but it has not been specifically tested.

Table 3.5 Essential oils containing high percentages of sesquiterpenes

Myrrh gum, *Commiphora myrrha* (headspace)	delta-elemene 28.7%	alpha-copaene 10%	beta-elemene 6.1%
Cedarwood (Texas), *Juniperus mexicana*	thujopsene 32%	alpha-cedrene 24.1%	cedrol 16%
Cedarwood (Virginia), *Juniperus virginiana*	cedrol 26%	alpha-cedrene 24.5%	thujopsene 15%
German Chamomile (Bulgaria), *Chamomilla recutita*	farnesene 27%	chamazulene 17%	alpha-bisabolol oxide B 11%
Ginger (China), *Zingiber officinale* Roscoe	ar-curcumene 16.3%	alpha-zingiberene 14.2%	beta-sesquiphellandrene 10.6%

Possible pheromonal effects

Farnesene is not only an essential oil constituent. In humans and other animals it is part of the chemical pathway to the formation of cholesterol and the steroid hormones (Lehninger, 1982).[18] In mice, as Ma et al. (1999) found, farnesene is present in the secretions of female mice which indicate the onset of estrus, and is responsible for inducing groups of female mice to go into estrus together.[19] Further research is needed to show whether farnesene plays a similar role in humans, and whether essential oils containing farnesene also have pheromonal effects.

Essential oils with high percentages of sesquiterpenes

Table 3.5 shows the three major constituents of several oils which contain high percentages of sesquiterpenes. The constituents which are not sesquiterpenes have been shaded grey.

Further reading

- For further information on solubility and other chemical reactions of monoterpenes and sesquiterpenes, this book is the six-volume 'bible'. University libraries are most likely to have copies: E. Guenther (1972), *The Essential Oils*, Krieger Press, Malabar, Flanders.
- For information on cosmetics and emulsions, there are several websites and books. Steffen Arctander's book is a classic: Steffen Arctander (1994) *Perfume and Flavor Materials of Natural Origin*, Allured Publishing Corp., Carol Stream, IL.

FUNCTIONAL GROUPS

In this chapter we consider terpenoid molecules with oxygen-containing functional groups attached to the monoterpene and sesquiterpene skeletons.

Molecules can have more than one functional group. An example is the phenol eugenol, which has a methyl ether functional group as well as a phenol functional group (see Phenols below).

As the functional group(s) of a molecule give it distinct properties, we can group molecules by their different functional groups. Generalisations can be made with caution about the therapeutic properties of molecules based on their functional group structure, but research is needed to discover the specific actions of each different constituent, alone and in synergy with the other constituents in the oil.

Alcohols

Terpenoid alcohols have a hydroxyl group attached to one of their carbon atoms. A hydroxyl group is also known as an –OH group, because it is exactly that: an oxygen atom bonded on the one hand to a carbon atom in the chain (represented by the –) and on the other hand to a hydrogen atom. Phenols are a type of alcohol, but due to their aromatic ring they are classed differently from the other types of alcohols. We will come to them later. To tell whether a diagram of a molecule represents an alcohol, look for an –OH group joined on to a carbon atom.

Monoterpenols

Structure

All monoterpenols have the monoterpenoid carbon chain of 10 carbon atoms and may contain a hydroxyl group anywhere along the chain. The addition of an –OH group usually means the number of double bonds reduces by one.

The –OH group can be added to carbon atoms at different places in the molecule, yielding primary, secondary and tertiary alcohols. In a primary alcohol the –OH group is attached to a carbon atom which in turn is attached to only one other carbon atom.

A secondary alcohol has two carbon atoms attached to the –OH carbon; a tertiary alcohol has three carbon atoms attached to that same carbon atom. This has relevance in terms of how easily the alcohol can be oxidised, which in turn affects its fate in the body in terms of how the liver will deal with it. It also has implications for the speed of degradation of the oil over time.

Primary alcohols are easily oxidised to aldehydes, secondary alcohols are oxidised to ketones, while tertiary alcohols are not oxidised at all. Only primary alcohols are easily metabolised and excreted by the body. Secondary and tertiary alcohols are likely to be more resistant to metabolism, and may stay in the body for longer (see Chapter 5).

Naming

Alcohols all have the ending -ol. The rest of their name may come from their parent monoterpene. An example is terpinen-4-ol which is derived from terpinene. The '4' indicates the position on the molecule where the –OH group is attached, calculated by counting clockwise round the ring, starting from the top and keeping the numbers as low as possible.

Solubility

Monoterpenols are mostly soluble in ethanol and are also soluble in other oils and non-polar solvents, because of their long carbon chain skeleton. Due to their polar group, they have slight solubility in water (0.1–0.4 g/l). Solubility details for all functional groups are taken from Guenther (1972).[1]

Volatility

Monoterpenols are often less volatile than monoterpenes, and their boiling points correspondingly higher. Analogous structures such as terpineol and limonene show the monoterpenol to be more viscous and less volatile. They are generally considered to be middle notes in perfumery.

Table 4.1 Examples of monoterpenols

Monoterpenol	Molecular structure
linalol (or linalool) $C_{10}H_{17}OH$ open chain (tertiary alcohol) Ho leaf 90% Rosewood 90% Basil (linalool) 40% Lavender 37%	
terpinen-4-ol $C_{10}H_{16}OH$ monocyclic (tertiary alcohol) Tea-tree 40% Sweet marjoram 25% Nutmeg 6%	
menthol $C_{10}H_{19}OH$ monocyclic (secondary alcohol) Cornmint 65% Peppermint 45%	
geraniol $C_{10}H_{17}OH$ open chain (primary alcohol) Palmarosa 75% Citronella 25% Geranium 20%	

Reactivity

When exposed to the air, primary monoterpenols slowly oxidise to aldehydes and acids, or form resins, which means that essential oils

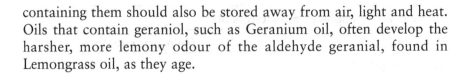

containing them should also be stored away from air, light and heat. Oils that contain geraniol, such as Geranium oil, often develop the harsher, more lemony odour of the aldehyde geranial, found in Lemongrass oil, as they age.

Toxic effects on the body

Monoterpenols do not exhibit any significant toxicity (Tisserand & Balacs, 1995) unless taken orally in amounts exceeding several milli-litres of oil.[2] Some can be mildly irritating to the skin, but no more so than monoterpenes.

Therapeutic effects of monoterpenols

Anti-infectious effects

Monoterpenols have been shown to have strong antibacterial and antifungal properties which, though not as strong as those of com-mercial disinfectants, have the advantage that they are usually mild on the skin and mucous membranes (Pénoël & Franchomme, 1990).[3] Pattnaik et al. (1997) investigated the activity of linalool, geraniol and menthol against bacteria and fungi. Linalool was most effective against bacteria and geraniol against fungi; menthol inhibited only half of the 18 bacteria and 12 fungi tested.[4] It would be interesting to see a comparison of all the monoterpenols found in essential oils against standard anti-microbial drugs.

Carson and Riley (1995) found that terpinen-4-ol, linalool and alpha-terpineol were active against *Escherichia coli*, *Staphylococcus aureus* and *Candida albicans*; terpinen-4-ol was the only constituent tested which was active against *Pseudomonas aeruginosa*.[5] Budhiraja et al. (1999) suggest that terpinen-4-ol, the major constituent of tea-tree oil (*Melaleuca alternifolia*), activates monocytes, the white blood cells involved in the immune response to infection.[6]

When you compare studies of the antibacterial activity of monoter-penes and monoterpenols, the monoterpenols all appear to be more active. One possible reason for this result may relate to the test method. Most antibacterial tests are done by the disc diffusion method in which the substance to be assessed (in this case the essential oil) is placed on a paper disc in the centre of a bacterially infected agar plate. The plate is covered and incubated, allowing bacteria to grow. After 24 hours, bacterial growth is assessed. Reduction in growth is proportional to the effectiveness of the test substance.

The trouble with using this method with essential oils is that the agar jelly the bacteria are growing on is made with water. Essential oil constituents that are more soluble in water (that is, monoterpenols with their –OH groups) might be expected to move more quickly through the jelly and thus reach a greater area of the plate than constituents with no –OH groups (the monoterpenes). If this process can be shown not to be affecting results, perhaps the monoterpenols do have a greater antibacterial effect than monoterpenes.

Analgesic properties

Menthol, linalool, alpha-terpineol and geraniol applied to the skin make the site of application feel cold. They also induce a slight local anaesthesia at the site of application (Pénoël & Franchomme, 1990).[7] Galeotti et al. (2002) tested the analgesic effects of (–)-menthol and (+)-menthol on mice. Oral and intravenous dosages of (–)-menthol between 3–10 mg/kg caused a dose-dependent increase in ability to tolerate pain (a hot plate test), but (+)-menthol did not have the same effects. Further investigation suggested that kappa-opioid receptors are affected by the (–)-menthol.[8]

Sedative properties

Linalool, a tertiary monoterpenol, exhibits some interesting sedative properties, even through inhalation. Buchbauer et al. (1991) showed that mice which had either Lavender oil (which contains about 40 per cent linalool), linalool or its ester linalyl acetate pumped into their cages all experienced a marked reduction in movement compared to the controls.[9] At this point, it is important to note that linalool occurs as two isomers (molecules with the same formula but a different spatial arrangement of atoms), (+)-linalool and (–)-linalool. Lavender contains only (–)-linalool. I have not found any references comparing the sedative effects of the two isomers. It is curious that Rosewood oil, which contains a mixture of both, has no tradition of being used as a sedative.

Elisabetsky et al. (1999) examined the effect of linalool (isomer unspecified) on the binding of L-[H3]-glutamate to neuronal membranes in the central nervous system. L-[H3]-glutamate is an excitatory neurotransmitter associated with convulsions. The results showed that linalool was an antagonist for glutamate *in vitro*, delayed aspartate-induced convulsions and blocked quinolinic acid-induced convulsions. This means that linalool may work as a sedative at the level of the central nervous system by modifying the response of neurons to L-[H3]-glutamate.[10]

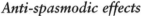

Anti-spasmodic effects

Lis-Balchin & Hart (1999) found that linalool and Lavender oil seemed to exert a spasmolytic effect on guinea pig intestinal smooth muscle. They suggest that the mechanism is through modulation of cyclic AMP (a cell signalling molecule) which affects the ability of smooth muscle to contract.[11]

Essential oils with high percentages of monoterpenols

Table 4.2 lists the main constituents of a number of high-monoterpenol essential oils. Note how common a constituent linalool is. Cells with grey shading do not show monoterpenols.

Sesquiterpenols

Structure

The sesquiterpenols are derived from sesquiterpenes by the addition of an –OH group; often both are present in the same oil. Different plants create special sesquiterpenols which become characteristic for that plant. Examples are patchoulol, which is only found in Patchouli oil, and the santalols found in Sandalwood oil. The structures often have two or three closed rings, which add to the complexity of the molecule. Oils that contain high proportions of sesquiterpenols are usually fairly viscous and slow-flowing. This impacts on the size of drops that emerge from a standard dropper-insert in an essential oil bottle, with one drop of Patchouli oil being much larger than a drop of Lemon oil, for example. Further discussion on drop size and dosage follows in Chapter 5.

Naming

Sesquiterpenols, like monoterpenols, all end in -ol, sometimes deriving the first part of their name from their parent sesquiterpene (for example, farnesene and farnesol).

Solubility

Sesquiterpenols are soluble in alcohol and vegetable oils. They are not soluble in water, in spite of the –OH group, because of the long carbon chain.

Table 4.2   Essential oils containing high percentages of monoterpenols

Basil (Portugal), *Ocimum basilicum*	linalool 38.2%	methyl chavicol 16.4%	beta-caryophyllene 7%
Rosewood, *Aniba rosaeodora*	linalool 85.3%	alpha-terpineol 3.5%	cis-linalool oxide 1.5%
Citronella (Ceylon), *Cymbopogon nardus* L. Rendle	geraniol 18%	limonene 9.7%	citronellol 8.4%
Geranium, bourbon, *Pelargonium graveolens*	citronellol 21.2%	geraniol 17.5%	linalool 12.9%
Ho leaf, *Cinnamomum camphora* Sieb. ssp. *formosana* var. *orientalis*	linalool 95%	camphor 0.4%	limonene 0.2%
Lavandin (abrial quality, France), *Lavandula hybrida* Rev.	linalool 33.5%	linalyl acetate 27.1%	camphor 9.5%
Sweet marjoram, *Marjorana hortensis* Moench (*Origanum majorana*)	terpinen-4-ol 36.3%	cis-sabinene hydrate 15.9%	para-cymene 9.5%
Neroli bigarade, *Citrus aurantium* var. *amara* flos	linalool 37.5%	limonene 16.6%	beta-pinene 11.8%
Palmarosa (India), *Cymbopogon martinii* Stapf. var. *motia*	geraniol 80%	geranyl acetate 8.3%	linalool 2.8%
Peppermint (USA), *Mentha piperita* L. var. Mitcham	menthol 42.8%	menthone 19.4%	sabinene hydrate 6.6%
Rose (Bulgaria), *Rosa damascena* Mill. (otto)	citronellol 33.4%	stearoptene waxes 24%	geraniol 18%
Rose (Egypt), *Rosa damascena* Mill.	2-phenyl ethyl alcohol 37.9%	geraniol 15.8%	citronellol 12.6%
Tea-tree, *Melaleuca alternifolia*	terpinen-4-ol 45.4%	gamma-terpinene 15.7%	alpha-terpinene 7.1%

Table 4.3   Examples of sesquiterpenols

Sesquiterpenol	Molecular structure
farnesol $C_{15}H_{25}OH$ open chain, unsaturated (primary alcohol) Jasmine, up to 10% Ylang ylang 2% *Rosa damascena* 1%	
viridiflorol $C_{15}H_{25}OH$ tricyclic, saturated (tertiary alcohol) Niaouli 18%	
patchoulol $C_{15}H_{24}OH$ tricyclic, saturated (tertiary alcohol) Patchouli 40%	
beta-santalol $C_{15}H_{24}OH$ bicyclic, unsaturated (primary alcohol) Sandalwood 20%	
Beta-eudesmol $C_{15}H_{25}OH$ bicyclic, unsaturated (tertiary alcohol) West Indian sandalwood, *Amyris balsamifera* 16% Ginger 0.9%	

Volatility

Again, volatility depends on boiling point, which in turn is dependent on structure and the internal attractions of the molecules for each other. The greater the attraction, the higher the boiling point and the less volatile the substance will be. Sesquiterpenols either have only a faint odour, or a woody, earthy odour. They are considered base notes in perfumery, as they evaporate slowly in comparison to the other types of essential oil molecules. The larger molecular size of sesquiterpenols also slows down their entry into the skin.

Reactivity

Similarly to monoterpenols, primary sesquiterpenols like farnesol will oxidise. This can have the effect of changing their odour and therapeutic properties. Secondary and tertiary sesquiterpenols are less reactive, and do not change in odour very much.

Toxic effects on the body

Sesquiterpenols seem to be relatively harmless, are usually listed as GRAS (generally recognised as safe) in toxicology indices and have even lower levels of irritation than monoterpenols.

Therapeutic effects of sesquiterpenols

Anti-inflammatory effects

Alpha-bisabolol, a sesquiterpenol found in German Chamomile oil, is even more anti-inflammatory than chamazulene.[12] A soothing cream called Camoderm or Camillosan contains Chamomile oil with bisabolol listed as an active ingredient. Several other sesquiterpenol-rich oils like Myrrh show anti-inflammatory properties, although more research is needed to show conclusively that it is the sesquiterpenols causing the effect.

Possible vascular effects

Luft et al. (1999) researched the effects of farnesol on smooth muscle cells of the rat aorta. They noted that farnesol blocks L-type calcium ion (Ca^{2+}) channels, thereby preventing contraction. When they fed farnesol to rats with hypertension (0.5g/kg doses), it significantly reduced the rats' blood pressure for up to 48 hours.[13] It may be that dietary farnesol would have similar effects in humans; topically applied farnesol is unlikely to have significant effect, as it may not even penetrate the epidermis due its molecular size.

Another group of sesquiterpenols, the cadinol isomers, also appeared to relax potassium ion (K^+) induced contractions of rat aortas, as Zygmunt et al. (1993) noted.[14] The effect was thought to be due to the blocking of calcium ion flow. Cadinol was much less powerful than nimodipine, a synthetic calcium antagonist. This reaction would lower the artificially raised blood pressure, but more research is definitely needed before recommending cadinol as a treatment for hypertension. T-cadinol constitutes 10.6 per cent of Hinoki wood oil.[15] Smaller amounts are found in *Artemisia alba* oil, while Black pepper and various pine species contain small amounts of other cadinol isomers (see Ready Reference List at the end of the book).

Neuronal effects

Alpha-eudesmol alleviates the over-stimulation of neuronal cells by blocking Ca^{2+} channels in neurons in the brain. Found in *Juniperus virginiana* (Cedarwood Virginian) oil, alpha-eudesmol particularly blocks those Ca^{2+} channels which are responsible for releasing glutamate during artificially induced ischaemic strokes in rats. Alpha-eudesmol was introduced into the intra-cerebroventricular fluid and significantly reduced the amount of brain damage and water volume in the brain after the brain injury (Asakura et al., 2000).[16]

Beta-eudesmol, found in *Amyris balsamifera* (West Indian Sandalwood) oil, alleviated electroshock convulsions in mice. It also reduced high level potassium-induced seizures and reduced the toxicity of organophosphate nerve poisoning. It may have potential as an antiepileptic substance, either on its own or in conjunction with phenytoin, an antiepileptic drug (Chiou et al., 1997).[17] It is found in some Chinese herbs and also in small amounts in Ginger and *Helichrysum italicum* (Everlasting or Immortelle) oils.

Anti-cancer effects

Rioja et al. (2000) examined the *in vitro* effects of farnesol on affected blood cells from patients with acute myeloid leukaemia. It seemed that farnesol selectively killed leukemic cells in preference to normal haemopoietic cells (cells which blood cells are derived from), at a level of 30 microM of farnesol.[18] Whether there will be similar effects *in vivo* remains to be seen. There are several isomers of farnesol and it wasn't specified which one was used in the study. Oils that contain various isomers of farnesol are *Helichrysum odoratissimum* (16 per cent), Cananga oil (21 per cent), and sometimes Jasmine (12 per cent) and Lemongrass (9 per cent). Farnesol isomers are also found in smaller amounts in oils of Ylang ylang and *Rosa damascena*.

Dietary nerolidol inhibited the growth of artificially induced neoplasms in the large intestine of male rats and reduced the numbers of tumours present at the end of the test period. Whether similar effects would be found in humans is not known, but Wattenberg (1991) suggests that nerolidol be investigated as an inhibitor of large intestine carcinogenesis.[19] The best source of nerolidol (about 90 per cent) is found in a chemotype of *Melaleuca quinquinervia* oil.[20] The other chemotype is the 1,8-cineole chemotype, which is commonly known as Niaouli oil. Several of the spice oils such as Cardamom and ambrette seed contain small amounts of nerolidol.

Anti-viral activity

Sandalwood oil is nearly all sesquiterpenes and sesquiterpenols, santalenes and santalols. Benencia and Courreges (1999) tested it against *Herpes simplex* viruses I and II *in vitro* and found that the oil inhibited replication of both viruses, but did not kill them. It was less effective at high viral load.[21] This may mean that Sandalwood oil could be used as a preventative for cold sores.

Possible anti-malarial activity

Lopes et al. (1999) found that nerolidol (see above under anti-cancer effects) completely inhibited the development of the malaria protozoan *in vitro*, seemingly by inhibition of glycoprotein synthesis.[22] The Amazon Waiapi Indians treat malaria by inhaling the vapour from a forest tree (*Virola surinamensis*) which contains nerolidol. There are other types of constituents which show definite anti-malarial activity, such as lactones from the *Artemisia* species.

Essential oils with high percentages of sesquiterpenols

The oils containing high percentages of sesquiterpenols (Table 4.4) are all viscous and woody-smelling. Notice that the constituents in the grey cells are all sesquiterpenes.

Table 4.4   Essential oils containing high percentages of sesquiterpenols

Cedarwood Virginia, *Juniperus virginiana*	cedrol 26%	alpha-cedrene 24.5%	thujopsene 15%
Patchouli (Indonesia), *Pogostemon cablin* Benth.	patchouli alcohol 33%	alpha-patchoulene 22%	beta-caryophyllene 20%
Sandalwood (India) *Santalum album*	cis-alpha-santalol 50%	cis-beta-santalol 20.9%	epi-beta-santalol 4.1%
Vetiver, *Vetivera zizanoides* Stapf.	vetiverol 50%	vetivenes 20%	alpha-vetivol 10%

Phenols

Structure

Phenols have a hydroxyl (–OH) group, attached to an aromatic or benzene ring (see Chapter 1). This is what distinguishes them from alcohols. The benzene ring is represented by a circle in the middle of the 6-membered ring because all the bonds in the ring between carbon atoms have equal value, 1.5 bonds. Sometimes it is drawn as a ring with three double bonds, but the circle representation is more accurate.

The aromatic ring has the effect of amplifying the electronegativity of the oxygen atom, so the hydrogen atom of the hydroxyl group is quite positive. The result of this is that phenols react like weak acids, and readily give up their hydrogen atoms to form ions. The net effect is that phenols are more antibacterial and also more irritant to the skin and mucous membranes than other molecules with –OH groups.

Naming

Because phenols contain an –OH group, they also have names ending -ol. The four described in Table 4.5 are the most common phenols in essential oils; nearly all other ol-ending names refer to ordinary alcohols.

Solubility

Due to the aromatic ring, which can disperse the excess negative charges of the oxygen atom, the hydrogen atom of the –OH group

Table 4.5 Examples of phenols

Phenol	Molecular structure
thymol $C_{10}H_{12}OH$ *Thymus vulgaris*, Thyme 40%	
carvacrol $C_{10}H_{12}OH$ *Origanum vulgaris*, Oregano 60%	
eugenol $C_9H_8OCH_3OH$ *Pimento dioica*, Allspice 80% Clove bud 70%	
chavicol C_9H_9OH *Pimenta racemosa*, West Indian bay 20%	

becomes very positive; this makes phenols into polar molecules which are, like the alcohols, more soluble in water than hydrocarbons. Phenols are soluble up to 1 g/l. For aromatherapy purposes, this still means that they are not very soluble in water, but they do dissolve in ethanol and other oils.

Volatility

Due to internal bonding via their hydroxyl (–OH) group, phenols are often found as crystals at room temperature (for example, thymol) and they have boiling points above 200°C. They do not evaporate very quickly, and tend to be pungent or very strong smelling.

Reactivity

Phenols are known as reactive molecules in chemistry, eliciting 'hazardous chemical' labels. They will also readily release their hydroxyl hydrogen atom and bind with other polar or positively charged molecules, including the protein molecules of the skin, thereby possibly causing damage to the normal structure of the skin.

Toxic effects on the body

Irritation

Phenols are the most irritant of the constituents to the skin and mucous membranes and can cause contact dermatitis and sensitisation dermatitis. I have tested a drop of neat carvacrol on my inner forearm, and found no noticeable effect. If I applied more, I dare say I would experience some warmth and irritation. The effect of neat carvacrol on the mucous membrane of the lips, however, is quite noticeable, and though not painful causes an unpleasant tingling.

Possible liver damage

Liver enzymes conjugate phenols with sulfonates and glucuronides that help solubilise the phenols for excretion. Some phenols, like eugenol, deplete the liver of glutathione, a detoxifying molecule, in the same way that paracetamol does.[23] Until more research shows otherwise, it is best to avoid the use of phenol-rich essential oils in cases where liver function is compromised.

Therapeutic effects of phenols

Rubefacient effect

Phenols are the most stimulating of the constituents, and can cause a rubefacient effect. 'Rubefacient' means causing a reddening of the skin (in fact, an irritation). The redness is due to the peripheral blood circulation being stimulated, so phenol-containing oils may be helpful in stimulating blood flow to cold hands and feet or stiff muscles. Different people have different sensitivities to oils containing phenols and the effects are dose-dependent (Pénoël & Franchomme, 1990).[24]

Anti-infectious effects

Phenols prevent most types of micro-organisms from growing and in most cases kill them, including *Staphylococcus aureus* and *Pseudomonas aeruginosa*. They are therefore useful for the treatment of acute infections such as boils, acne, vaginal thrush and other diseases of that nature, though care must be taken with the concentration so as not to cause irritation or sensitisation (Pénoël & Franchomme, 1990).[25]

Consentino et al. (1999) researched the effects of three types of Thyme oil on various bacteria and confirmed that it was the high phenolic content which caused these effects.[26] Ultee et al. (1999) researched the bactericidal mechanism of carvacrol in *Bacillus cereus* cultures and found that carvacrol appears to interact with the cell membranes, causing them to increase in permeability to potassium and hydrogen ions. This increase in permeability causes a gradual impairment of bacterial cell processes and leads to cell death.[27]

Cholesterol lowering effects

Case et al. (1995) found that dietary supplements of thymol and carvacrol (1 mmol/kg) significantly reduced the serum cholesterol levels of cockerels. They suggested that it was due to the ability of the molecules to interact with the enzymes required for cholesterol manufacture by cells.[28] However, for a 70 kg person the dose needed would be about 10.5g orally per day, which would most likely cause gastrointestinal irritation. It is probably unlikely that topically applied thymol and carvacrol would reach the bloodstream in sufficient quantities to have significant effect.

Essential oils with high percentages of phenols

In Table 4.6, only the constituents in the white cells are phenols. Note the presence of gamma-terpinene and para-cymene in the grey cells. It is likely that they are precursors of phenols in the biosynthetic pathway in the plant.

Table 4.6   Essential oils containing high percentages of phenols

Ajowan, *Trachyspermum copticum* L. Link	thymol 61%	para-cymene 15.5%	gamma-terpinene 11%
Bay (West Indian), *Pimenta racemosa* Mill. JS Moore	eugenol 56%	chavicol 21.6%	myrcene 13%
Cinnamon leaf, *Cinnamomum zeylanicum* Blume	eugenol 87%	benzyl benzoate 2.6%	beta-caryophyllene 1.8%
Clove bud (Madagascar), *Eugenia caryophyllus* (Spreng.) Bullock	eugenol 76.6%	beta-caryophyllene 9.8%	eugenyl acetate 7.6%
Oregano (Greece) *Origanum vulgare* ssp. *viride* (Boiss.) Hayak (thymol chemotype)	thymol 85.6%	carvacrol 4.3%	gamma-terpinene 2.7%
Savory, summer (Italy), *Satureja hortensis*	carvacrol 48%	gamma-terpinene 28%	para-cymene 7%
Thyme (Italy), *Thymus vulgaris*	thymol 27.4%	para-cymene 21.9%	gamma-terpinene 12%

Aldehydes

Structure

Aldehydes are characterised by a carbonyl group on a terminal carbon atom. A carbonyl group is an oxygen atom double-bonded to a carbon atom (C=O). The fourth bond is always a hydrogen atom. They are derived from primary alcohols by a process called oxidation.

Naming

Aldehydes will either be called by their common name, followed by the word 'aldehyde', for example cumin aldehyde; or they will end in

Table 4.7   Examples of aldehydes

Aldehyde	Molecular structure
trans-2-hexenal C_5H_9CHO non-terpenoid Sweet marjoram 0.3% Clary sage 0.05%	
citronellal $C_9H_{17}CHO$ monoterpenoid *Eucalyptus citriodora* 75% Citronella 35% *Litsea cubeba*, May chang 10%	
neral $C_9H_{15}CHO$ monoterpenoid Melissa 35% Lemongrass 35% *Litsea cubeba* 30% Ginger 3% Lemon 1%	
geranial $C_9H_{15}CHO$ monoterpenoid Lemongrass 50% *Litsea cubeba* 25–40% Melissa 20% Ginger 10% Lemon 2%	

-al, for example cuminal. Aldehydes are derived from parent primary alcohols, for example, citronellal from citronellol. Table 4.7 shows a non-terpenoid aldehyde, trans-2-hexenal, and three monoterpenoid aldehydes, but it is possible also to get sesquiterpenoid aldehydes. Sandalwood oils (*Santalum album* and *Santalum spicatum*) have santalals, for example.

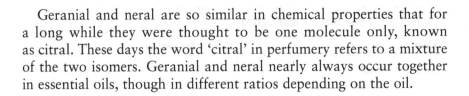

Geranial and neral are so similar in chemical properties that for a long while they were thought to be one molecule only, known as citral. These days the word 'citral' in perfumery refers to a mixture of the two isomers. Geranial and neral nearly always occur together in essential oils, though in different ratios depending on the oil.

Solubility

Carbonyl groups are polar, making low molecular weight aldehydes slightly soluble in water. They are also soluble in ethanol and other oils. Sesquiterpenoid aldehydes are not soluble in water though they are soluble in ethanol and other oils.

Volatility

Monoterpenoid aldehydes are mostly as volatile as monoterpenols. Most aldehydes smell stronger than their equivalent alcohol, and the non-terpenoid aldehydes octanal, nonanal and decanal in citrus oils can be detected at very low concentrations by the human nose.

Reactivity

Aldehydes are unstable and oxidise to the acid in the presence of oxygen.

Toxic effects on the body

As citral is a commonly used aldehyde in the flavour and fragrance industry, so more is known about its possible negative effects. In general I would caution the dermal use of oils with high percentages of any aldehyde. In practice, the oils to watch out for are Lemongrass and other lemon-scented oils like May Chang and Lemon eucalyptus, and Cinnamon bark oil. Fortunately the citrus oils contain only small percentages of non-terpenoid aldehydes, but beware—even these small percentages may be enough to cause mucous membrane irritation.

Mucous membrane irritants
Aldehydes are often mucous membrane irritants and will sometimes cause the eyes to water, like onions do. Citral is irritating to the mucous membranes, as I found out once when I tried a gargle with

one drop of Lemongrass oil in an attempt to cure a sore throat. The pain from the irritation of the citral was enough to make me forget about the original pain, but it is not an experience I recommend.

Glaucoma

Tisserand and Balacs (1995) cite some cautions for use of citral-rich oils, in particular for glaucoma sufferers, as very low oral dosages (2–5 micrograms) per day in monkeys caused an increase in ocular pressure.[29]

Hyperplasia or pre-cancerous conditions

Other studies, reporting a possibly malign effect of citral, showed it causing overgrowth (hyperplasia) of prostatic and vaginal cells in rats after topical application at human-equivalent doses of about 10 ml.[30] This may have significance for women who have an abnormal Pap smear, although the only circumstance I could imagine where there would be any real risk of stimulating cell-growth would be if you were applying it topically, which is unlikely.

Liver enzyme effects

Cinnamaldehyde, found in Cinnamon oils, depresses rat liver glutathione levels, which may interfere with paracetamol (acetaminophen) metabolism and cause liver toxicity if used at the same time (Tisserand & Balacs, 1995).[31] This is possibly due to the benzene ring.

Therapeutic effects of aldehydes

Possible calming properties

Melissa oil, with both citronellal and citral, is traditionally thought of as calming and sedative. Pénoël and Franchomme (1990) suggest this is due to the aldehyde content.[32]

Anti-infectious agents

The carbonyl group appears to increase the antibacterial effect of a molecule in a similar way to the –OH group of the alcohols. Cinnamaldehyde from Cinnamon oil has anti-infectious properties on par with the phenols and is also a skin and mucous membrane irritant, probably due to its benzene ring.

One use for aldehyde-containing oils could be in the management of opportunistic fungal infections in the last stages of AIDS, as

Viollon and Chaumon (1994) suggest. The constituents that showed particular efficacy against *Cryptococcus neoformans* were the phenols thymol and carvacrol and the aldehydes citral and cinnamaldehyde.[33]

Anti-melanoma activity

Trans-2-hexenal, found in low concentrations in many of the leaf oils such as Oregano, Thyme and Sweet Marjoram, was shown by Iersal et al. (1996) to inhibit glutathione-S-transferase enzymes in melanoma cells.[34] Glutathione is a molecule used by all cells as an anti-oxidant; if its production is inhibited, melanoma cells are more likely to be killed by chemotherapy drugs.

Essential oils with high percentages of aldehydes

In Table 4.8, following the previous style, the constituents in the white cells are aldehydes. Constituents in the grey cells are not aldehydes.

Table 4.8 Essential oils containing high percentages of aldehydes			
Cinnamon bark, *Cinnamomum zeylanicum*	cinnamaldehyde 74%	eugenol 8.8%	cinnamyl acetate 5.1%
Citronella (Java), *Cymbopogon winterianus*	citronellal 36.8%	geraniol 21.4%	citronellol 15%
Lemon balm/ Melissa, *Melissa officinalis*	geranial 45%	neral 35%	6-methyl-5-hepten -2-one 3%
May chang, *Litsea cubeba* (berry)	geranial 40%	neral 33.8%	limonene 8.3%

Ketones

Structure

Ketones have a carbonyl group ($C=O$) like the aldehydes, but always on a carbon atom which is bonded to two other carbon atoms. They

Table 4.9   Examples of ketones

Ketone	Molecular structure
menthone $C_{10}H_{18}O$ monocyclic Peppermint 30% Geranium, bourbon 2%	
camphor $C_{10}H_{16}O$ bicyclic Rosemary 15–30% Sage 22% Yarrow 12% Spike lavender 15%	
thujone $C_{10}H_{16}O$ bicyclic Thuja/Cedarleaf 45% Sage 20–40% Wormwood 20% Tansy, *Tanacetum vulgare* 60%	

are derived from secondary alcohols by oxidation. The oxidation of ketones usually takes longer than the oxidation of aldehydes and mostly takes place in the plant.

Naming

Ketones usually have the ending '-one'. An exception is camphor, though it is also known as 2-bornanone. The name 'camphor' has been used for centuries, which is why it persists today. Ketones are derived from secondary alcohols, for example, menthone from menthol. There are both monoterpenoid and sesquiterpenoid ketones, though monoterpenoid ketones are more common.

Solubility

The polar carbonyl group present in ketones makes them slightly soluble in water. They are also soluble in ethanol and other oils.

Volatility

Ketones have fairly high boiling points and can occur in crystalline form at room temperature (for example, camphor). Monoterpenoid ketones with a closed ring seem to share a minty-camphoraceous odour which feels quite 'penetrating' when you smell it.

Reactivity

Ketones are relatively stable and may pose problems in the body because they are resistant to metabolism by the liver.

Toxic effects on the body

Several ketones have toxic effects on the body at quite low oral dosages (e.g. camphor, about 3g in an adult human (Tisserand and Balacs, 1995, p. 186)). Dermal application is unlikely to yield problematic blood levels on single doses, though there is a possibility of build-up with repeated daily dosages due to the long half-life of terpenoids in the body (see Chapter 5).

Neurotoxic effects

Some terpenoid ketones are known to produce convulsions and liver and central nervous system (CNS) damage. According to Tisserand and Balacs (1995), the ketones with potential convulsant or CNS-damaging properties to be aware of are:

- artemisia ketone in Wormwood oil;
- thujone in Mugwort, Dalmatian sage (*Salvia officinalis*), Tansy, Thuja, Wormwood and Western red cedar (*Thuja plicata*) oils;
- pulegone in Buchu and Pennyroyal oils;
- camphor in Mugwort, Ho leaf (camphor CT), Dalmatian sage and possibly Rosemary (camphor CT);
- pinocamphone and iso-pinocamphone in Hyssop oil.[35]

Alpha-thujone blocks gamma-amino butyric acid receptors in the brain, causing an over-excitation of neurons. In rats, the oral lethal

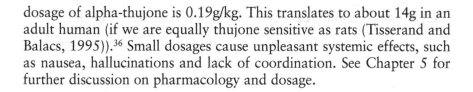

dosage of alpha-thujone is 0.19g/kg. This translates to about 14g in an adult human (if we are equally thujone sensitive as rats (Tisserand and Balacs, 1995)).[36] Small dosages cause unpleasant systemic effects, such as nausea, hallucinations and lack of coordination. See Chapter 5 for further discussion on pharmacology and dosage.

Therapeutic effects of ketones

Effects on mucosal secretions
Ketones are thought to be effective at reducing mucosal secretions caused by both respiratory and genito-urinary infections (Pénoël & Franchomme, 1990).[37] Further research is required to confirm which ketones are most effective.

Wound-healing properties
Pénoël and Franchomme (1990) recommend the use of oils high in ketones for wounds, scars, burns and surgical wounds, suggesting that the ketones prevent cheloid formation and over-production of scar tissue.[38]

Anti-haematomal properties
Constituents containing two ketone groups (known as diketones) are supposed to have anti-haematomal properties, in particular the italidiones of the oil of *Helichrysum italicum* ssp. *serotinum* (Pénoël & Franchomme, 1990).[39] Anecdotal reports from colleagues confirm this suggestion, however, I have not found any supporting references.

Venous effects
Camphor has been shown to cause vasodilation of dog and rat portal veins by relaxation of the venous smooth muscle at dosages of 0.6–6 mg/kg of camphor in 0.1 ml of a carrier applied to the skin. The initial reason for the research was to determine whether camphor could alleviate haemorrhoidal bleeding and inflammation, and the results suggest that perhaps it could be used for such a purpose, though with caution due to dermal sensitisation (Xie et al., 1992).[40]

Anti-viral properties
Apparently papilloma virus, herpes viruses and other viruses which attack the nervous system, such as shingles, can be arrested by ketones (Pénoël & Franchomme, 1990).[41] This remains to be confirmed in further scientific research.

Table 4.10   Essential oils containing high percentages of ketones

Artemisia alba (Belgium)	iso-pinocamphone 34.6	camphor 21.1%	1,8-cineole 5.7%
Annual wormwood (Yugoslavia), Artemisia alba	artemisia ketone 44.8%	1,8-cineole 9.6%	camphor 6.3%
Buchu, Barosma betulina Bertl. (also Agathosma betulina)	(−)-menthone 35%	diosphenol 12%	(−)-pulegone 11%
Camphor (Japan), Cinnamomum camphora (yellow or brown camphor)	camphor 51.5%	safrole 13.4%	1,8-cineole 4.75%
Caraway (Netherlands), Carum carvi L.	(+)-carvone 50%	limonene 46%	cis-dihyrdocarvone 0.48%
Cedarleaf (Canada), Thuja occidentalis	alpha-thujone 56%	fenchone 15%	beta-thujone 14.7%
Hyssop (France), Hyssopus officinalis	iso-pinocamphone 32.6%	beta-pinene 22.9%	pinocamphone 12.2%
Spearmint (Greece), Mentha spicata	(−)-carvone 42.8%	dihydrocarvone 15.7%	1,8-cineole 5.8%
Pennyroyal, Mentha pulegium	(+)-pulegone 63.5%	(+)-isomenthone 19.7%	(+)-neoisomenthol 5.7%
Rosemary (Spain), Rosmarinus officinalis (camphor chemotype)	alpha-pinene 22%	camphor 17%	1,8-cineole 17%
Rue (Egypt), Ruta graveolens	2-undecanone 49.2%	2-nonanone 24.7%	2-nonyl acetate 6.2%
Sage (Dalmatian), Salvia officinalis	alpha-thujone 37.1%	beta-thujone 14.2%	camphor 12.3%
Tansy (Belgium), Tanacetum vulgare	beta-thujone 50%	trans-chrysanthemyl acetate 20%	camphor 6.4%

Turmeric (Indonesia), *Curcuma longa*	turmerone 29.5%	ar-turmerone 24.7%	turmerol 20%
Mugwort (Germany), *Artemisia absinthum*	beta-thujone 46%	sabinyl acetate 25%	trans-sabinol 3.2%
Yarrow, *Achillea millefolium*	camphor 17.7%	sabinene 12.3%	1,8-cineole 9.5%

Essential oils with high percentages of ketones

In Table 4.10, the cells shaded grey are not ketones. Notice the number of oils that have more than one type of ketone, and that the oxide 1,8-cineole accompanies many of the ketones.

Acids and esters

The acids found in plants are known as carboxylic acids. They are mostly non-terpenoid, with carbon chains of two to five atoms, which allows them to dissolve in water. Most plant acids are so soluble in water that they are extracted into the waters of distillation, known as hydrosols. Consequently, they do not occur in high proportions in the essential oils. The most common carboxylic acid is acetic or ethanoic acid, which we know as vinegar (see Figure 4.1).

Some resins from which essential oils are extracted contain terpenoid

Figure 4.1 Structure of the carboxylic acid, acetic acid (also known as ethanoic acid)

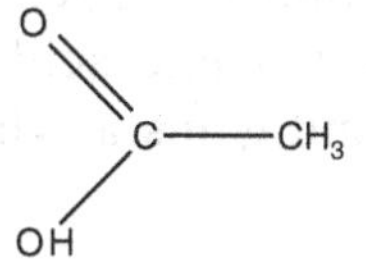

carboxylic acids, like the anti-inflammatory tri-terpenoid boswellic acids in frankincense (*Boswellia* species) resin. These do not make it through into the steam-distilled essential oils. Solvents such as supercritical CO_2 can extract higher proportions of boswellic acids as the solvent extraction method does not rely on volatility of the constituents.

In the plant, acids are easily combined with alcohols to make

Figure 4.2   The esterification reaction of linalool and acetic acid leading to formation of an ester

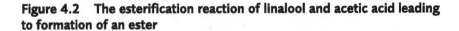

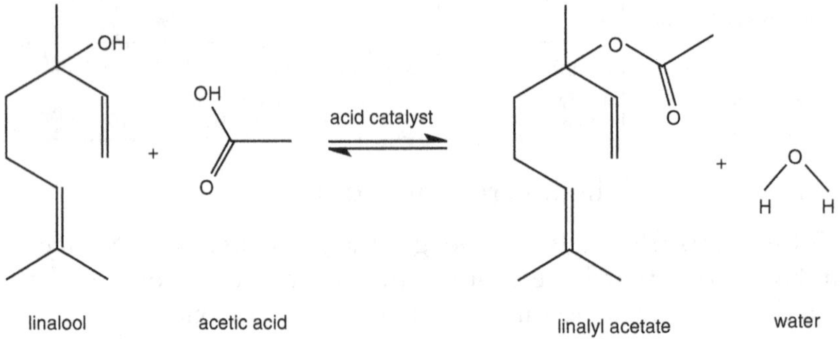

esters. The acid hydroxyl group will readily leave if there are catalyst hydrogen ions present. The positively charged acid remnant will then displace the hydrogen of the alcohol hydroxyl group to produce the ester and a molecule of water. This reaction is reversible, given the right combination of water and acids. It is one of the sources of variation between the odour of the living plant material and the steam-distilled essential oil. Esters are often formed or broken down during steam distillation. Figure 4.2 shows the formation of the ester linalyl acetate from linalool and acetic acid.

Naming

As esters are made from alcohols and acids and are named after both parent molecules, 'linalool + acetic acid' becomes 'linalyl acetate'. The alcohol drops the -ol and gains a -yl; the acid drops the -ic and gains an -ate.

Solubility

Terpenoid acids are the most soluble in water of the essential oil constituents, due to their polar functional groups, but are still not very soluble (about 2 g/l). Esters are not very soluble, because the potentially polar part of their structure is neutralised by the bulk of the two non-polar carbon chains.

Table 4.11 Examples of esters

Ester	Molecular structure
linalyl acetate $C_{10}H_{17}OCOCH_3$ Clary sage 50% Lavender 40% Bergamot 25%	
benzyl benzoate $C_7H_8OCOC_7H_8$ Narcissus absolute 20% Jasmine absolute 16% Ylang ylang 7%	
isobutyl angelate $C_4H_9OCOC_4H_7$ Roman chamomile 35%	

Volatility

Again, volatility depends on structure and boiling point, but on the whole, acids and esters are not significantly more volatile than their equivalent alcohols. Esters can be detected at relatively low concentrations, so they affect the odour quality of an oil significantly.

Reactivity

Esters are generally quite stable, particularly in essential oils where there is no water available for hydrolysis. However, it is still possible that the double bonds can oxidise as for the terpenes. They are generally recognised as safe.

Toxic effects on the body

As the acids are easily metabolised by the body and excreted in the urine, they do not generally pose a toxicity problem. As with all terpenoid molecules, esters and acids can cause skin sensitisation when used over long periods, and esters are held responsible for the drying effects on the skin in oils such as Lavender, according to Pénoël and Franchomme (1990).[42]

The esters with known toxicity are methyl salicylate and sabinyl acetate. Methyl salicylate is found in Wintergreen oil. Oral dosages as low as 4 ml in children can cause respiratory failure and death (Tisserand & Balacs, 1995).[43] Sabinyl acetate is embryotoxic and causes liver toxicity in mice (Pages et al., 1989).[44] The only oil likely to be used in aromatherapy that can contain sabinyl acetate is Spanish sage (*Salvia lavandulaefolia*).

Therapeutic effects of acids and esters

Anti-spasmodic and sedative properties
Pénoël and Franchomme (1990) suggest these are the most important properties of esters. Esters are thought to regulate and re-equilibrate the sympathetic nervous system and neuro-endocrine system.[45]

Buchbauer et al. (1991) experimented on mice with linalool, linalyl acetate and Lavender oil, and found that both linalool and linalyl acetate exhibited a sedative effect on the mice, seeming to depress the motor cortex particularly.[46]

Pénoël and Franchomme mention an anecdote where *Anthemis nobilis* (Roman chamomile) oil applied to the neck was effective as a pre-operative relaxing agent. Roman chamomile oil contains large amounts of non-terpenoid esters. The main constituent is isobutyl angelate (35 per cent), which they suggest is one of the 'grand anti-spasmodics of the pharmacopoeia'.[47] Further research is needed to confirm this suggestion.

Anti-inflammatory and analgesic effects
Methyl salicylate, found in Wintergreen oil (greater than 98 per cent), is thought to have mild analgesic, counter-irritant and anti-inflammatory effects similar to aspirin, acetyl salicylate. However, some individuals show extreme sensitivity to methyl salicylate (personal observation), so it should be used with caution and patch testing prior to use.

Essential oils with high percentages of esters

The grey cells in Table 4.12 are not esters. Note the presence of linalool with linalyl acetate in the oils.

Table 4.12   Essential oils containing high percentages of esters

Clary sage (USA), *Salvia sclarea*	linalyl acetate 49%	linalool 24%	germacrene D 3%
Lavandin (abrial quality, France), *Lavandula hybrida* Rev.	linalool 33.5%	linalyl acetate 27.1%	camphor 9.5%
Lavender (France), *Lavandula angustifolia* Mill.	linalyl acetate 40%	linalool 31.5%	(Z)-beta-ocimene 6.7%
Myrtle (Spain, wild growing), *Myrtus communis* L.	myrtenyl acetate 35.9%	1,8-cineole 29.9%	alpha-pinene 8.1%
Petitgrain bigarade, *Citrus aurantium* L. ssp. *amara*	linalyl acetate 45.5%	linalool 24.1%	alpha-terpineol 5.2%
Roman chamomile (Japan), *Anthemis nobilis*	isobutyl angelate 35.9%	2-methylbutyl angelate 15.3%	methallyl angelate 8.7%
Wintergreen (China), *Gaultheria procumbens*	methyl salicylate 90%	safrole 5%	linalool 2%

Phenyl methyl ethers

Structure

Most ethers occurring in the essential oils are phenolic ethers. They are derived from the hydroxyl group of the phenol (or alcohol), the hydrogen of which is replaced with a short carbon chain, either a methyl group ($-CH_3$), or a two carbon ethyl group ($-CH_2-CH_3$). The electronegativity of the oxygen atom causes opposite dipoles from the adjacent carbon atoms, so the net polarity of ethers is slight.

Table 4.13 Examples of phenyl methyl ethers	
Phenol methyl ether	**Molecular structure**
methyl chavicol (chavicol methyl ether, methoxy allyl benzene) $C_9H_9OCH_3$ Basil (Comoro Islands) 85% Tarragon, *Artemisia dranunculus* 60% Fennel 5% Star anise, *Illicium verum* 5%	
eugenol $C_9H_9OHOCH_3$ Clove bud 75% Pimento 40% Basil (Portugal) 5% *Rosa damascena* 1%	
trans-anethole $C_9H_9OCH_3$ Aniseed, *Pimpinella anisum* 93% Sweet fennel 70% Star anise, *Illicium verum* 70%	

Naming

Phenyl methyl ethers are derived from phenols where the hydrogen atom of the –OH group has been exchanged for a carbon chain (usually a methyl group, –CH$_3$). They are often named after the phenol, plus the carbon chain group name and with the word 'ether' attached, for example, chavicol methyl ether. 'Chavicol' is the name of the phenol, 'methyl' indicates a –CH$_3$ group and 'ether' indicates that they are linked by an oxygen group. In French and other languages, it is not uncommon to see the abbreviation 'chavicol M.E.'.

Another way of naming ethers is by using the carbon chain name (methyl, ethyl, propyl, butyl, etc.) and adding '-oxy' to indicate the presence of the oxygen group, for example, methoxy allyl benzene.

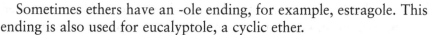

Sometimes ethers have an -ole ending, for example, estragole. This ending is also used for eucalyptole, a cyclic ether.

Note that eugenol is both a phenol and a methyl ether. It takes the -ol ending like a phenol, but shares properties of both functional groups. The methyl ether group seems to modulate the irritancy of the phenolic –OH group.

Solubility

Most ethers are soluble in ethanol, but negligibly soluble in water. The polarity of the oxygen atom is shielded by the –CH$_3$ (methyl) group.

Volatility

The ethers tend to have boiling points below 200°C and are thus fairly volatile compared to their cousins the lactones and coumarins, most of which are solids at room temperature. Again this is dependent on the structure of the molecules and the extent to which they experience inter-molecular attraction. Ethers tend to have strong smells and can be detected at quite low percentages in an oil.

Reactivity

Ethers are not very reactive when exposed to heat and light, as both the methyl group and the benzene ring are resistant to oxidation. The only changes which could occur are changes to the C=C bond in the 3-carbon (propanoid) 'tail' of the molecule.

Toxic effects on the body

Stupefying or psychotropic effects

In large oral doses, ethers can have stupefying effects, leading to convulsions and even death. Tisserand and Balacs (1995) note that myristicin and elemicin, two ethers found in Nutmeg oil, are supposed to be psychotropic if ingested in sufficient quantities, but that they are fairly toxic at the doses which will produce the psychotropic effect.[48] Myristicin appears to inhibit the monoamine oxidase enzymes in the brain and increase the levels of brain serotonin, both of which would give rise to euphoric effects in humans. Both myristicin and elemicin have more than one ether group, which appears to intensify their psychotropic effects. However, Tisserand and Balacs also noted that trans-anethole had psychotropic effects in mice at an oral dosage level equivalent to 20 ml in humans. It is unlikely that dermally applied ether-rich oils could cause psychotropic effects in the dosages generally used in aromatherapy.

Liver toxicity

The ether safrole, found in Yellow and Brown camphor (*Cinnamomum camphora*) oils, is noted for its liver carcinogenic effects in animals given daily oral doses over a long period of time (Tisserand & Balacs, 1995).[49] As with constituents such as the ketone pulegone, the metabolites of the compound are more carcinogenic than the original compound. The presence of the benzene ring also makes the molecules more difficult to break down in the liver, meaning they stay in the body for longer.

Therapeutic effects of phenyl methyl ethers

The effects discussed in this section are so far only known to apply to the phenyl methyl ethers, for example, chavicol methyl ether and anethole. Other types of ethers, such as the spiroethers found in Clary sage oil, have not yet been researched.

Antispasmodic effects

Ethers appear to have an antispasmodic effect on the smooth muscles of the intestines and the genito-urinary tract (Pénoël & Franchomme, 1990).[50] Mills and Bone (2000) cite several references that support the smooth muscle antispasmodic activity of Fennel oil, presumably due to its anethole content.[51] Albuquerque et al. (1995) found that both anethole and methyl chavicol (estragole) block skeletal muscle contraction when applied to artificially stimulated isolated frog muscle tissue.[52]

Anaesthetic effects

Ghelardini et al. (2001) found that trans-anethole blocks generation of isolated rat phrenic nerve impulses at dosages of 1 microgram/ml, but that eugenol had no effect. They suggest that this demonstrates a possible anaesthetic effect.[53]

Anti-infectious properties

The phenyl methyl ethers have an 'all or nothing' effect against micro-organisms, but most are effective against some type of micro-organism (Pénoël & Franchomme, 1990).[54] Eugenol, one of the main constituents of Clove oil, is used in dental mouthwashes. Seltzer (1992) reviewed the use of eugenol in dentistry and found it to be antibacterial and analgesic, with anti-inflammatory properties at low dosages but risk of irritancy at higher dosages.[55]

Essential oils with high percentages of phenyl methyl ethers

One oil not mentioned in Table 4.14 is Ylang ylang (*Cananga odorata*), which contains the ether known as p-cresyl methyl ether. The ether content varies depending on the grade of Ylang ylang, occurring up to 16 per cent in the most expensive grades. Therapeutic effects of p-cresyl methyl ether have not yet been determined.

Again, constituents in the grey boxes are not ethers.

Table 4.14 Essential oils containing high percentages of phenyl methyl ethers

Anise (Spain), *Pimpinella anisum*	(E)-anethole 96%	limonene 0.6%	anisaldehyde 0.56%
Basil (Comoro Islands)	methyl chavicol 85%	1,8-cineole 3.25%	para-cymene 2.7%
Sweet fennel (Turkey), *Foeniculum vulgare* Mill. var. *dulce*	(E)-anethole 80%	limonene 6%	methyl chavicol 4.5%
Star anise (China), *Illicium verum* Hook. F.	(E)-anethole 71.5%	foeniculin 14.5% (an azulene)	methyl chavicol 5%
Tarragon (USA), *Artemisia dranunculus*	methyl chavicol 80%	beta-ocimene 14%	limonene 2.5%

Cyclic ethers or oxides

Cyclic ethers are similar to phenyl methyl ethers but the oxygen atom is included in a ring formation rather than being sandwiched by a short carbon chain. The most ubiquitous in essential oils is 1,8-cineole, or eucalyptole as it is commonly known. Several cyclic ethers are known as oxides, for example linalool oxide, caryophyllene oxide and rose oxide. Cyclic ethers or oxides often contribute significantly to the odour of a steam-distilled essential oil, but not much is known about their therapeutic effects.

Structure

Cyclic ethers are formed where an –OH group is modified by the removal of the H atoms and the oxygen atom bonds to another carbon atom to form a closed ring. Oxides that form during steam distillation can have many different forms, and it can be quite difficult to assess which structures are present in an oil.

Naming

Eucalyptole (1,8-cineole) follows the -ole ending for ethers. Other constituents keep the name of the molecule they were derived from, followed by 'oxide', for example linalool oxide. If the closed ether ring has six atoms, it can be called a 'pyran' ring. If it has five atoms (including the O atom) it can be called a 'furan' ring. Menthofuran is an example of a furan found in *Mentha piperita* oil. Some metabolites of constituents have an even smaller 3-membered ether ring known as an 'epoxy' ring. Figure 4.3 shows furan and epoxy ether rings.

Peroxides are rare in essential oils. They have two oxygen atoms next to each other, C-O-O-C. The only well-known one is ascaridole—a toxic cyclic peroxide from Wormseed oil (see Chapter 8).

Figure 4.3 Structure of epoxy ether and furan rings, showing placement of oxygen atom within carbon chain

epoxy ether   furan

Table 4.15 Examples of cyclic ethers and oxides

Cyclic ether/Oxide	Molecular structure
1,8-cineole (eucalyptole) $C_{10}H_{18}O$ bicyclic *Eucalyptus globulus* 70% Cardamom 30% Spike lavender 15% Sage 15% Rosemary 15%	
menthofuran $C_{10}H_{14}O$ bicyclic, furan ring Peppermint 4% Cornmint, *Mentha arvensis* 1%	
rose oxide $C_{10}H_{18}O$ monocyclic, pyran ring Geranium 1% Cistus 0.5% *Rosa damascena* 0.3%	

Solubility

Cyclic ethers and oxides are negligibly soluble in water but dissolve well in ethanol and other oils.

Volatility

Monoterpenoid cyclic ethers and oxides are probably as volatile as monoterpenols. Of the functional groups, they are possibly the strongest odorants, giving characteristic odours to the oils in which they occur, even at percentages as low as 0.3% (for example, rose oxide in *Rosa damascena* oil). Sesquiterpenoid oxides have similar volatility to sesquiterpenols.

Reactivity

Oxides tend to decompose into alcohols under conditions of heat and light; they can also form long chain polymers of terpenoid molecules which end up forming a sticky residue. This is because oxides, in particular epoxides, exist in very strained conformations and readily react to form a less strained structure. Epoxy resin, a preparation used in sealing wood, works by reacting with oxygen in the air to form a solid layer of polymerised epoxide.

Toxic effects on the body

Neurotoxic effects

Little research has been done on the oxides, with the exception of 1,8-cineole, which has similar neurotropic properties as the ethers and has been known to cause ataxia, slurred speech, unconsciousness and convulsions (Day et al., 1997; Burkhard et al., 1999).[56] Darben et al. (1998) cite a case where a six-year-old child experienced such symptoms after topical application of a home remedy for urticaria which contained Eucalyptus oil.[57]

Ascaridole, a terpenoid peroxide found in Wormseed (*Chenopodium ambrosiodes*) and Boldo (*Peumus boldus*) oils, is neurotoxic to humans. Doses of one or two drops used to be given for the treatment of intestinal worms, but even at these dosages could cause vomiting, nausea and other signs of toxicity (Pénoël & Franchomme, 1990).[58] Wormseed and Boldo oils should not be used in aromatherapy.

Liver toxic effects

Menthofuran, found in trace amounts in Peppermint and other mint oils, is also a metabolite of pulegone, a toxic ketone found in Pennyroyal oil. It destroys liver cytochrome P450 enzymes responsible for detoxifying foreign substances, and as such can be very damaging to the liver (Tisserand & Balacs, 1995).[59]

Respiratory irritation

1,8-cineole has a strong odour with a definite component of mucous membrane irritation. Pénoël and Franchomme (1990) suggest that care must be taken when using oils like *Eucalyptus globulus* in asthma, as irritation of mucous membranes can trigger an asthma attack.[60]

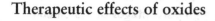

Therapeutic effects of oxides

What follows is largely an account of the effects of 1,8-cineole, as most other oxides have not yet been researched for therapeutic effects.

Expectorant effect

Mucous glands and cilia of the respiratory tract are thought to be stimulated by oxides, in particular linalool oxide and 1,8-cineole, which results in an expectorant effect. Research by Ulmer & Schott (1991) supports the suggestion that 1,8-cineole has an expectorant effect. They conducted a double blind study of an inhalation product called Gelomyrtol forte containing 1,8-cineole, with patients with chronic obstructive bronchitis. Results showed that the product decreased coughing, increased ease of expectoration and helped lengthen the breath in the experimental group as compared to the control. The colour of the sputum also improved over the 14-day treatment period.[61]

Anti-inflammatory effect in asthma

Juergens et al. (1998) suggest that 1,8-cineole may exhibit anti-inflammatory effects in bronchial asthma by inhibiting the leukotriene B4 and prostaglandin E2 pathways in blood monocytes of humans with bronchial asthma. The effect was noted in monocytes of healthy volunteers as well as the asthma sufferers.[62]

Essential oils with high percentages of cyclic ethers 1,8-cineole

As ascaridole is the only other oxide that occurs in high percentages in two essential oils, Table 4.16 is a list of oils that contain 1,8-cineole. The two chemotypes of Rosemary oil are shown to contrast the different levels of cineole. The grey cells are obviously not oxides or cyclic ethers.

Lactones

Structure

Lactones are probably derived from an intramolecular esterification reaction. This can take place when a molecule has an acid at one end of the chain and an alcohol group either three or four carbons down

Table 4.16 Essential oils containing high percentages of 1,8-cineole			
Eucalyptus globulus (Spain)	1,8-cineole 66.1%	alpha-pinene 14.7%	limonene 3%
Cajeput, Melaleuca leucadendron L.	1,8-cineole 41.1%	alpha-terpineol 8.7%	para-cymene 6.8%
Cardamom (Reunion), Elettaria cardamomum L. Maton var. alpha-minor	1,8-cineole 48.4%	alpha-terpinyl acetate 24%	limonene 6%
Niaouli (Madagascar), Melaleuca quinquenervia Cav.	1,8-cineole 41.8%	viridiflorol 18.1%	limonene 5.5%
Myrtle (Spain, wild growing), Myrtus communis L.	myrtenyl acetate 35.9%	1,8-cineole 29.9%	alpha-pinene 8.1%
Rosemary (Tunisia), cineole chemotype	1,8-cineole 51.3%	camphor 10.6%	alpha-pinene 10%
Rosemary (Spain), camphor chemotype	alpha-pinene	camphor 17%	1,8-cineole 17%
Spike lavender (Spain), Lavandula latifolia Medicus	1,8-cineole 36.3%	linalool 30.3%	camphor 8%
Angelica (root), Angelica archangelica	alpha-pinene 25%	1,8-cineole 14.5%	alpha-phellandrene 13.5%

the chain (Morrison & Boyd, 1987, p. 874). Lactones can be either monoterpenoid or sesquiterpenoid and always have a carbonyl group (C=O) next to an oxygen atom that is part of a closed ring.

Naming

Lactones are usually known by their common names, as the chemical ones are too lengthy. They tend to end in -lactone, but can also end in -in, or -ine. Examples of lactones appear in Table 4.17.

Table 4.17   Examples of lactones

Lactone	Molecular structure
helenaline $C_{15}H_{18}O_4$ sesquiterpenoid tricyclic *Arnica montana* (in alcoholic extract)	
nepetalactone $C_{10}H_{14}O_2$ monoterpenoid bicyclic Catnip 90%, *Nepeta cataria*	
alantolactone $C_{15}H_{20}O_2$ sesquiterpenoid tricyclic Sweet inula, *Inula graveolens* 20% Elecampane, *Inula helenium* 20%	
delta-jasmine lactone $C_{10}H_{16}O_2$ non-terpenoid, monocyclic pyran ring Jasmine 1%	

Sweet Inula and Catnip oils are not used widely in aromatherapy, and there is no commercially available oil of Arnica. However, these lactones are worth considering as there may be other lactones in oils yet to be discovered which may share similar properties. There are many isomers of jasmine lactone, which are found in Jasmine oil at about 1 per cent.

Solubility

Lactones are negligibly soluble in water, like most other essential oil constituents. Sesquiterpenoid lactones are even less soluble.

Volatility

Monoterpenoid lactones are more volatile than sesquiterpenoid ones, as might be expected. Sesquiterpenoid lactones are not very volatile and their often earthy and pungent odours linger for a long time.

Reactivity

Lactones have similar reactive properties to ethers and ketones and do not easily oxidise. They are also not likely to be metabolised easily and probably stay for longer in the body.

Toxic effects on the body

Skin allergies and sensitisation

The main caution about oils with high percentages of lactones is the likelihood of their causing skin allergies, sensitisation and blistering. Daisy-like plants from the Compositae (Asteraceae) family often have sesquiterpenoid lactones which are responsible for a large bulk of the allergic contact dermatitis among florists. Chrysanthemums are particularly noted for their allergenic effects (Gordon, 1999).[63] It is interesting to note that alantolactone, found in Sweet Inula oil and cited as a mucolytic by Pénoël and Franchomme (1990), is also among those listed as a sensitising lactone by Tisserand and Balacs (1995, p. 182). Fortunately, lactones occur only in small amounts in a few essential oils.

Some people have reported allergies to both Roman Chamomile and German Chamomile oils that suggest a lactone-type reaction (McGeorge & Steele, 1981; Van Ketel, 1982).[64] There are traces of lactones in the steam-distilled oils, and there could possibly be more in super-critical CO_2 extracts.

Interference with liver metabolism of drugs

Helenalin, a sesquiterpenoid lactone found in aqueous extracts of *Arnica montana*, reduces liver enzyme activity in rats (Joydnis-Liebert et al., 2000).[65] Other sesquiterpenoid lactones also have this effect. In particular, the glutathione-S-transferase and NADPH-cytochrome P450 reductase enzymes are affected. These two enzymes are responsible for keeping glutathione and cytochrome P450 levels high enough to protect the liver from toxic substances, including free radicals and drugs like paracetamol (Tisserand & Balacs, 2000).[66]

Therapeutic effects of lactones

Possible expectorant effects

Pénoël and Franchomme (1990) suggest that alantolactone and iso-alantolactone, found in *Inula graveolens* (Sweet Inula) and *Inula helenium* (Elecampane) oils, are the active components when these oils are used in the treatment of chronic catarrh and bronchial congestion.[67] They report a case study where the inhalation of Sweet Inula oil caused a 'healing crisis' but within two days completely cleared a chronic bronchial and sinus infection in an eight-year-old girl.

Anti-malarial effects

Essential oil from the Chinese herb Qinghao, *Artemisia annua* (also known as Sweet Annie), contains small amounts of the sesquiterpenoid lactones called artemisinins, which show promise of being active against the malaria parasite. The herbal tincture contains more artemisinins than the essential oil due to their low volatility. Unfortunately, the oil also contains artemisia ketone, which Pénoël and Franchomme (1990) suggest is neurotoxic in large doses,[68] so I would not recommend use of the oil in aromatherapy until further research has been carried out.

Anti-inflammatory effects

In spite of some lactones being allergens for some people, other research is showing that sesquiterpenoid lactones have powerful anti-inflammatory properties. Mazor et al. (2000) studied the anti-inflammatory effects of isohelenin (also known as iso-alantolactone) and parthenolide in human respiratory epithelium.[69] They found that the lactones inhibited the expression of the gene for interleukin-8, a molecule which promotes the inflammatory process, and suggest that this may be one mechanism whereby the lactones exert their anti-inflammatory effect. Perhaps the results Pénoël and Franchomme noted with the use of *Inula graveolens* oil were due more to the anti-inflammatory effect of iso-alantolactone than to the expectorant effects of alantolactone.

Mills and Bone (2000) summarise several papers on the anti-inflammatory activity of helenalin from arnica extracts.[70] Lyss et al. (1997) found helenalin had a similar anti-inflammatory effect to that of isohelenin (iso-alantolactone).[71] The effect is due to inhibition of the transcription factor NF-kappaB, a major controlling factor in the inflammatory process. This is a different mechanism than that used by aspirin and other non-steroidal anti-inflammatory drugs.

Possible analgesic effects

Aydin et al. (1998) examined nepetalactone found in *Nepeta caesarea* (catnip genus) essential oil.[72] They suggest that oral dosages provide not only marked sedation in rats, but also mediated the rats' response to mechanical pain. This is not the same as having pain in the first place and using an analgesic to stop it, but nevertheless is of some interest. They also suggest that nepetalactone has a specificity for particular opioid receptors, which would put it in the same category of analgesics as codeine.

Essential oils with high percentages of lactones

No essential oils used in aromatherapy have lactones as their major constituents. I show Catnip in Table 4.18 only because there may be some future interest in using it if the analgesic effects of nepetalactone in rats translate to humans. Note that the two isomers of nepetalactone make up about 90 per cent of the oil.

Table 4.18 An essential oil containing high percentages of lactones			
Catnip, *Nepeta cataria*	nepetalactone–1 80.5%	nepetalactone–2 10%	beta-caryophyllene 4.4%

Coumarins

Structure

Coumarins have a lactone ring adjoined to a benzene ring, which in turn can have several different functional groups attached. It is difficult to predict which functional group will dominate the properties of the molecule.

Naming

Coumarins are usually known by their common names, as the chemical ones are too lengthy (see Table 4.19). They tend to end with -in, for example, herniarin, but can also end in -one, for example, umbelliferone, otherwise known as 7-hydroxy-coumarin. Some researchers refer to coumarins as benzo-alpha-pyrones, where the -pyrone ending indicates the presence of a pyran ring with a ketone attached onto the ring.

Table 4.19   Examples of coumarins

Coumarin	Molecular structure
coumarin $C_9H_6O_2$ bicyclic, non-terpenoid Tonka bean 50% Cassia 12% *Narcissus poeticus* absolute 6%	
herniarin (7-methoxy-coumarin) $C_9H_6(OCH_3)O_2$ bicyclic, non-terpenoid Tarragon, *Artemisia dranunculus* 0.2% Lavender (trace)	
umbelliferone (7-hydroxycoumarin) $C_9H_6(OH)O_2$ bicyclic, non-terpenoid Dill, *Anethum graveolens* 2%	

Furanocoumarins (also known as psoralens), like bergaptene, have a 5-membered furan ring attached to the coumarin structure (see in Chapter 8).

Solubility

Coumarins are less soluble in ethanol than most other constituents and are negligibly soluble in water.

Volatility

Coumarins tend to be solids at room temperature. Again this is dependent on the structure of the molecules and the extent to which they experience inter-molecular attraction. Coumarins are more prevalent in solvent-extracted oils as they are not volatile enough to be extracted in high quantities by steam distillation.

Reactivity

Coumarins do not readily oxidise. They are also not metabolised particularly easily, due to the benzene ring and the resistance of the lactone ring to oxidation.

Toxic effects on the body

Drug interaction with anticoagulants

Dicoumarol, a molecule with double coumarin groups, is a potent anticoagulant. It is generated in the stomachs of herbivores that have eaten too much Sweet Clover which contains coumarin and causes abnormal internal bleeding (Lehninger, 1982).[73] Other coumarins may also interfere with blood-clotting, although this has not been proven conclusively. Until further research proves otherwise, it is as well to avoid oils with coumarins if a patient is undergoing warfarin treatment for a thrombosis. Oils from the Umbelliferae family, such as Angelica, Celery Seed, Dill, Fennel and Parsley, are likely to contain various coumarins, although in small amounts. Note the coumarin group present in the structure of warfarin in Figure 4.4.

Figure 4.4   Structure of warfarin, showing similarity to coumarin

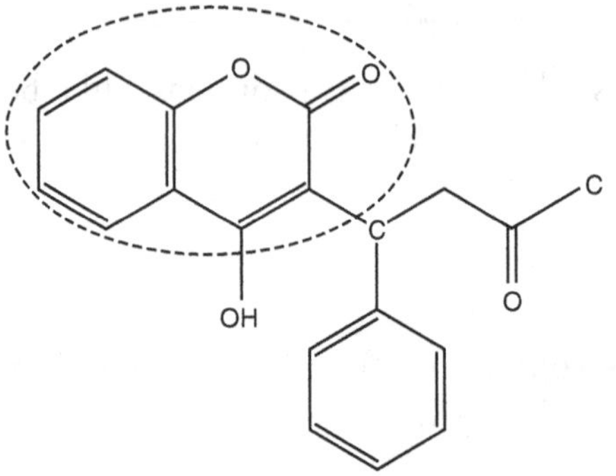

Phototoxicity

The well-known UV-sensitising effect of Bergamot oil on the skin is due to the presence of a furanocoumarin, bergapten or bergaptene, which links up with the DNA of cells responsible for the manufacture of melanin. It does this because its chemical structure can align with the pyrimidine bases in the DNA molecules.

Once linked, the structure of bergaptene (benzene ring next to an unsaturated furan ring) allows for the production of an oxygen free

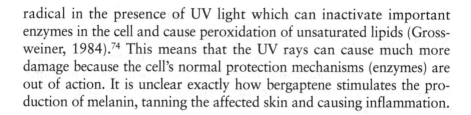

radical in the presence of UV light which can inactivate important enzymes in the cell and cause peroxidation of unsaturated lipids (Grossweiner, 1984).[74] This means that the UV rays can cause much more damage because the cell's normal protection mechanisms (enzymes) are out of action. It is unclear exactly how bergaptene stimulates the production of melanin, tanning the affected skin and causing inflammation.

Therapeutic effects of coumarins

Sedative effects
According to Pénoël and Franchomme (1990), coumarins can have hypnotic effects in large doses and are known as sleep-inducers.[75] I have not found other research to support this, but it may well be worth exploring.

Antilymphoedemic effects
Casley-Smith (1999) reviewed 50 trials of different coumarins used in the treatment of lymphoedema, some of which were oral dosages, others topical.[76] Oral coumarin treatment significantly reduced oedema, by up to 55 per cent, as compared to control groups. Trials that went over several months had greater effects than shorter trials. Topically applied coumarins also reduced oedema, but not as effectively. The only side-effect noted was idiosyncratic hepatitis (3 in 1000 likelihood).

Burgos et al. (1999) reported use of coumarin in 90 mg/day oral dosages with women suffering from lymphoedema as a result of surgical removal of breast cancer.[77] During 12 months of coumarin therapy the oedema significantly improved, with the volume of the arms decreasing and the heaviness, hardness and discomfort in the nerves reducing as well.

Cardiovascular effects
Mills and Bone (2000) report the results of several studies on the effects of coumarin on various problems associated with ischaemic heart disease (blocked coronary arteries).[78] They suggest that coumarin may be helpful in managing this disease.

Antispasmodic effects
Some coumarins have antispasmodic effects, for example khellin in Khella oil (*Ammi visnaga*) and coumarins from *Angelica* species used in herbal medicine.

Essential oils with high percentages of coumarins

No oils used in aromatherapy have high percentages of coumarins. The only product that comes close is an extract from the tonka bean (see Table 4.20). Crystalline coumarin (where the structure gets its name from) can be collected from the harvested beans, but no essential oil is commercially produced.

Table 4.20 An essential oil containing high percentages of coumarins			
Tonka bean, Dipteryx odorata	coumarin about 50%	Not listed	Not listed

Summary of hazardous and therapeutic properties of functional groups

The main purpose of trying to assign therapeutic properties to functional groups is to make it easier to grasp the chemistry. It would be nice if there were some general principles, and to that end I have summarised the information in Chapters 3 and 4 on the constituents in Table 4.21. However, I think it is also useful to examine each molecule in its own right, as molecules within a functional group often have different individual properties. Question marks (?) indicate possible action unsupported by research evidence.

Further reading

- For details on essential oil toxicity see R. Tisserand and T. Balacs (eds) (1995), *Essential Oil Safety: A Guide for Health Care Professionals*, Churchill Livingstone, Edinburgh.
- For an excellent book with details on other types of molecules found in therapeutic herbs and information on essential oil-rich herbs see S. Mills and K. Bone (eds) (2002), *Principles and Practice of Phytotherapy: Modern Herbal Medicine*, Churchill Livingstone, Edinburgh.
- For an expensive but comprehensive book on essential oil chemistry from a perfumery perspective, see D.G. Williams (1996), *The Chemistry of Essential Oils*, Micelle Press, Weymouth, Dorset, England. The book has 'scratch 'n' sniff' panels, which are fun.

Table 4.21   Summary of therapeutic properties of essential oil constituents

Functional group	Hazards	Therapeutic properties	Constituent
Monoterpene	Skin and mucous membrane irritating if oxidised (pinenes and delta-3-carene)	Dissolve gallstones	Limonene
		Tumour prevention	Limonene
		Relief of muscular aches	Pinene, para-cymene
Sesquiterpene	Skin and mucous membrane irritating if oxidised	Anti-inflammatory	Most: beta-caryophyllene, chamazulene
		Anti-tumoral	Elemene
Monoterpenol		Anti-infectious	Most: terpinen-4-ol, geraniol, linalool, alpha-terpineol
		Analgesic	Geraniol, linalool, alpha-terpineol, menthol
		Sedative	(-)-linalool
		Anti-spasmodic	linalool
Sesquiterpenol		Anti-inflammatory	alpha-bisabolol; santalols
		Arterial relaxant	Farnesol; cadinols
		Neuronal blocking effects	Eudesmols
		Anti-cancer	Farnesol; nerolidol
		Viral inhibitor (*Herpes* spp.)	Santalols
		Anti-malarial	Nerolidol
Phenol	Skin and mucous membrane irritation (most); possible liver damage (most)	Rubefacient	Most
		Anti-infectious	Most: carvacrol; eugenol
		Cholesterol lowering	Thymol, carvacrol
Aldehyde	Skin and mucous membrane irritation (most); glaucoma (citral); hyperplasia (citral); liver enzyme depletion (cinnamaldehyde)	Anti-infectious	Cinnamaldehyde, citral
		Anti-melanoma	Trans-2-hexenal

Table 4.21 *continued*

Ketone	Neurotoxic effects (most, except carvone)	Mucolytic	Most (?)
		Wound-healing	Most (?)
		Anti-haematomal	Italidiones
		Vasodilation	Camphor
		Anti-viral	Most (?)
Ester		Antispasmodic and sedative	Linalyl acetate; isobutyl angelate
		Anti-inflammatory and analgesic	Methyl salicylate
Ether	Psychotropic (myristicine, elemecin, anethole); liver toxicity (safrole, most)	Anti-spasmodic	Trans-anethole, methyl chavicol
		Anaesthetic	Trans-anethole
Oxide	Neurotoxic (1,8-cineole); liver toxicity (1,8-cineole); respiratory irritation (1,8-cineole)	Expectorant	1,8-cineole
		Anti-inflammatory effect	1,8-cineole
Lactone	Skin allergies and sensitisation (most); depletion of liver enzymes (most)	Mucolytic and expectorant	Alantolactone
		Anti-malarial	Artemisinins
		Anti-inflammatory	Isoalantolactone, helenalin
		Analgesic	Nepetalactone
Coumarin	Anti-coagulant (hydroxycoumarins); phototoxicity (bergaptene, furocoumarins, psoralens)	Sedative	Coumarin (?)
		Anti-lymphoedema	Coumarin, umbelliferone
		Cardiovascular	Coumarin
		Antispasmodic	Khellin

THE PHARMACOLOGY OF ESSENTIAL OILS

Pharmacology is the study of the actions and effects of drugs on living organisms. According to Bryant et al. (2003), the World Health Organization defines a drug as 'any substance or product that is used or intended to be used to modify or explore physiological systems or pathological states for the benefit of the recipient.'[1] This includes essential oils.

There are two main branches of pharmacology: pharmacokinetics (what the body does to the drug) and pharmacodynamics (what the drug does to the body). We will define useful concepts from both areas first, then the pharmacokinetics of essential oils followed by pharmacodynamic interactions of essential oil constituents with relevance to disease states of various body organs and tissues. (See Further reading for suggested pharmacology references.)

Pharmacodynamics

Pharmacodynamics (Greek, *pharmaco*-, drug-related, -*dynamic*, mechanism of action) is concerned with the following: how drugs bind to drug targets (usually by a protein of some kind); biochemical, physiological and possible adverse effects of drug binding; potency, specificity and efficacy of drugs; drug interactions with food, herbs and other drugs. Drug binding affinity for the target site, selectivity, potency and maximal efficacy are all required to determine the usefulness of a drug, and are defined below.

All drug molecules work by interacting with target molecules in the body. Essential oil molecules interact with various types of target molecules or target sites: cell membranes; neuronal and muscular ion channels; neurotransmitter receptors; G-protein-coupled receptors and second messengers; intra-cellular and extra-cellular enzymes; and in the case of phototoxic furocoumarins, even with DNA molecules.

Binding affinity and specificity

Binding affinity is the extent to which a molecule binds with another molecule. For most therapeutic interactions, the binding affinity between drug and target molecule is higher than that between drug and non-target molecules (for example, plasma proteins).

Agonists and antagonists

A drug is an agonist if it binds to the same receptor and has the same effect as the endogenous (produced by the body) molecule. A drug is an antagonist if it binds to the same receptor, but has either no effect, or the opposite effect to the endogenous molecule. For example, nicotine is an agonist for acetylcholine receptors, because it causes the same stimulation of memory formation as acetylcholine (the endogenous molecule).

Selectivity

The biomedical model of pharmaceutical medicine has as a central myth the 'lock and key' concept of drug action, that is, a drug (key) activates or blocks a single target (lock). In practice, drugs bind to a multitude of sites and each target is usually composed of a family of sites, for example, receptor subtypes. A useful drug has some selectivity and avoids binding to sites that result in toxicity. As essential oil constituents commonly interfere with basic membrane physiology (see below), they are often characterised as having low selectivity (i.e., they interact with a number of different types of target sites). However, evidence is now accumulating that the diverse compounds found in essential oils do exert potent, selective effects. Furthermore, because essential oils contain several different chemicals they potentially have a wide spectrum of effects which may merge with the symptoms present in some disease states. For example, people with nervous anxiety are often prone to dermatitis and eczema. An essential oil such as Lavender can reduce anxiety at a central nervous system level, and is also a topical anti-inflammatory agent.

Drug potency and efficacy

The potency of a drug is determined by the amount that is required to give the desired therapeutic effect. Potency is usually calculated from experiments where the subject is given increasing doses until undesirable or toxic effects are observed. The results of these experiments are plotted in graphs known as 'concentration–response curves'. A potent drug achieves the therapeutic effect at low dosages (usually in the range of 1×10^{-9}g per ml). However, potent drugs may also pose risks of toxicity (see Figure 5.6).

Essential oil constituents vary in their potency and efficacy. Two groups of researchers, Galeotti et al. (2001) and Ghelardini et al. (1999) compared the *in vitro* anaesthetic properties of different essential oil constituents.[2] Table 5.1 summarises the results.

Table 5.1   Comparison of potency and efficacy of some essential oil constituents as local anaesthetics

Constituent/drug	Approximate dosage that caused 50% anaesthesia	Minimum dosage at which contractions were totally prevented	Comment	Authors
Procaine	1×10^{-4} μg/ml	1 μg/ml		Galeotti et al. (2000)
(+)-menthol	2.5×10^{-4} μg/ml	1 μg/ml	Equally effective and potent as procaine for total anaesthesia of nerve	Galeotti et al. (2000)
(−)-menthol	9×10^{-4} μg/ml	1 μg/ml	Equally effective and potent as procaine and (+)-menthol for total anaesthesia of nerve	Galeotti et al. (2000)
(−)-menthone	N/A	N/A	Did not achieve more than 25% anaesthesia, even at dosage of 1 μg/ml	Galeotti et al. (2000)
Thymol	N/A	N/A	Did not achieve more than 25% anaesthesia, even at dosage of 1 μg/ml	Galeotti et al. (2000)

Table 5.1	*continued*			
Procaine	1×10^{-3} µg/ml	0.1 µg/ml	10-fold differences in results with other research	Ghelardini et al. (1999)
Lidocaine	1×10^{-3} µg/ml	0.1 µg/ml	As potent as procaine	Ghelardini et al. (1999)
Lavandula angustifolia oil	9×10^{-2} µg/ml	1000 µg/ml	Equally as effective as procaine and menthols, but not as potent	Ghelardini et al. (1999)
Linalyl acetate	9×10^{-2} µg/ml	1000 µg/ml	Equally as effective as procaine and menthols	Ghelardini et al. (1999)
Linalool	1.1 µg/ml	1000 µg/ml	Equally as effective as procaine and menthols	Ghelardini et al. (1999)
Citrus reticulata (Mandarin) oil	N/A	N/A	Did not achieve more than about 15% anaesthesia, even at dosage of 1000 µg/ml	Ghelardini et al. (1999)
Citrus limon (Lemon) oil	N/A	N/A	Did not achieve more than about 15% anaesthesia, even at dosage of 1000 µg/ml	Ghelardini et al. (1999)

These results suggest that menthol is more potent than Lavender oil, linalyl acetate and linalool as an anaesthetic for the phrenic nerve-hemidiaphragm *in vitro*. Of course, we must be cautious about extrapolating these results to other types of nerves, and to living animals or humans.

Plasma concentration

Plasma concentration is often used to measure how much of a drug is present in the body at any one time (measured in either ng or mg of substance per ml of plasma). However, it is not necessarily an accurate measure of the total amount of a drug in the body. Drugs that have a high capacity to associate with target tissues have low plasma concentrations. Plasma concentration–time profiles can be created by measuring the plasma concentrations of a drug over time.

Figure 5.1 Typical plasma concentration–time curve of a therapeutic dose of a hypothetical terpenoid compound

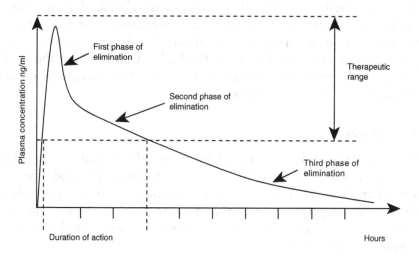

Concentration measurements are usually taken until the drug plasma concentration has reduced to zero. Figure 5.1 shows a plasma concentration–time graph of a therapeutic range of a drug.

The area under the curve can be calculated and, as long as the drug shows a simple distribution (as here), the area under the curve is a good estimate of the proportion of the initial dose that was actually absorbed into the body.

Half-life

The half-life of a drug is defined as the length of time taken for its plasma concentration to fall by one half of initial measurement. Essential oil molecules with high lipophilicity can show what is known as bi- or tri-phasic distribution, with an initial fast reduction followed by a more steady tailing off. The three phases are shown in Figure 5.1. Each phase has a different half-life, with the slow phases usually having a much longer half-life. Half-life is a composite measure, combining the rates of absorption, distribution, formation of metabolites and elimination rates. It can be modified by changes in function of metabolic and excretory organs like the liver and the kidneys.

Pharmacokinetics

Pharmacokinetics (Greek, *pharmaco-*, drug-related; *-kinetic*, movement) is concerned with absorption, distribution, metabolism and excretion of drugs. The concentration of a drug at its target site and the length of time it stays in the body (and therefore the likelihood of its having the desired therapeutic effect) depends on the different rates of these processes.

Absorption

The two main application routes for essential oils that can be safely used by aromatherapists with no medical training are inhalation and dermal application. These deliver topical dosages of essential oils to the skin or respiratory membranes. Essential oil constituents are absorbed into the circulatory system at varying speeds and in varying amounts for each method.

Absorption of essential oils via the respiratory system
The lungs provide a large absorption surface area, with a thin epithelial layer and lipophilic absorbing surface. They also have a high blood flow which increases the rate of absorption into the blood stream. Absorption rates depend to a large extent on molecular size, polarity and solubility. The monoterpenoids are likely to be absorbed more rapidly than sesquiterpenoids as their smaller size allows them to diffuse more rapidly.

Given these parameters, inhalation is more likely to rapidly deliver a dose of essential oils into the circulation than dermal application. To enhance absorption via inhalation, it would help to concentrate the essential oil vapour in the inhaled air over a period of time, perhaps using an asthma-medication delivery system.

Falk et al. (1990) noted that in human volunteers exposed to 450mg/m³ of alpha-pinene in the air for two hours (similar to working conditions in a pine saw-mill), uptake of alpha-pinene from the air was about 60 per cent.[3] Similar results were found for d-limonene at the same air concentration, except that uptake was about 70 per cent (A. Falk-Filipsson et al., 1993).[4] These experiments had very high levels of an essential oil constituent in the air compared with an average aromatherapy vaporisation of three drops of Rosemary oil (drop size approx. 26 mg) vaporised in a 30m³ room yielding about 2.6 mg/m³ (Svoboda et al., 2000).[5] It is not known what percentage of vaporised oils would be absorbed from the air in a typical aromatherapy treatment room, as some of the oil is absorbed by fabric surfaces in the room (e.g. curtains, towels, carpet). Unfortunately the mouse fragrance inhalation experiments done by researchers at the Institute of Pharmaceutical Chemistry, University of Vienna, did not calculate the percentage uptake of fragrance compounds, and the wood shavings in the mouse cages would probably have absorbed a large proportion of the fragrance compounds, thereby reducing the dosage.[6] However, perhaps the experiments more realistically reflect environmental conditions in an aromatherapy treatment situation (wood shavings make great carpet!).

Römmelt et al. (1978) exposed human volunteers to 10 per cent aqueous solutions of alpha-pinene, limonene, camphor and borneol sprayed into the air. Amounts absorbed were 61 per cent, 66 per cent, 54 per cent and 58 per cent respectively, but when they measured blood concentrations they found only 4.6–7.5 per cent of the expected absorbed constituents in the blood.[7] This suggests that either the constituents are metabolised in the mucus layer of the respiratory tract, or that they are very rapidly absorbed into tissues from the blood stream.

Absorption of essential oils via the skin

Essential oils applied dermally are either mixed with vegetable oil or applied undiluted. Vegetable oil retards evaporation of constituents and may enhance their penetration into the skin. The essential oil mixture is left on the skin for several hours, allowing a slow, continuous absorption compared to the more rapid absorption via the lungs. According to Bryant et al. (2003), massage increases the blood flow to the skin, and is therefore thought to increase the rate of absorption.[8] Essential oil constituents penetrate the skin layers at different rates, though several monoterpenoids were found to reach their peak plasma concentrations (between 0.05ng/ml and 9ng/ml) within ten minutes of application of 2 grams of Pinimenthol cream (as reviewed by Kohlert et al., 2000).[9]

Pénoël and Franchomme (1990) also agree that dermally applied essential oil constituents can be detected in the blood within a few minutes of application.[10] Jäger et al. (1992) found that after massaging 1.5 grams of a 2 per cent lavender oil blend (equivalent to doses of 7.23 mg linalool and 8.64 mg linalyl acetate) into a volunteer's stomach skin (surface area 376 cm^2) for 10 minutes, linalool and linalyl acetate could be detected in the blood after 5 minutes at a plasma concentration of about 30 ng/ml.[11] Peak plasma concentration was 121.08 ng/ml (18.76 mins) for linalool, and 100.17 ng/ml (19.55 mins) for linalyl acetate. The area under the curve was not calculated, which would have given an idea of the proportion of the initial dose absorbed into the blood stream.

The different binding affinities of essential oil constituents for plasma lipoproteins and different tissues in the body will determine their plasma concentration and whether they show multiple pharmacokinetic phases. Essential oil constituents are likely to be rapidly compartmentalised into the lipid-rich tissues of the body, such as the brain and adipose layers.

Figure 5.2   Schematic representation of the layers of the skin and underlying structures

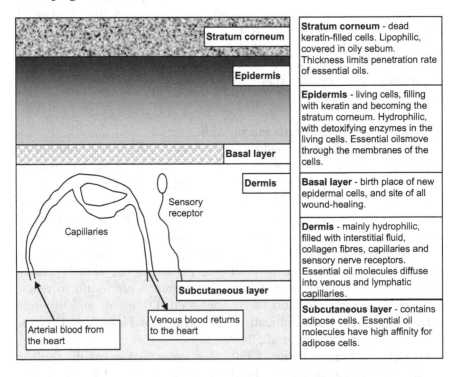

Stratum corneum	**Stratum corneum** - dead keratin-filled cells. Lipophilic, covered in oily sebum. Thickness limits penetration rate of essential oils.
Epidermis	**Epidermis** - living cells, filling with keratin and becoming the stratum corneum. Hydrophilic, with detoxifying enzymes in the living cells. Essential oils move through the membranes of the cells.
Basal layer	**Basal layer** - birth place of new epidermal cells, and site of all wound-healing.
Dermis Sensory receptor Capillaries	**Dermis** - mainly hydrophilic, filled with interstitial fluid, collagen fibres, capillaries and sensory nerve receptors. Essential oil molecules diffuse into venous and lymphatic capillaries.
Subcutaneous layer Arterial blood from the heart Venous blood returns to the heart	**Subcutaneous layer** - contains adipose cells. Essential oil molecules have high affinity for adipose cells.

Figure 5.2 shows the layers of the skin, the epidermis, dermis and subcutaneous adipose layer. Terpenoid constituents readily diffuse into the protective sebaceous layer that covers the skin, due to their

lipophilic nature. Terpenoid constituents can also dissolve sebum, and if applied neat to the skin can remove the sebaceous layer, thus drying the skin. Williams and Barry (1991) suggested that terpenes affect the lipids between the cells of the stratum corneum, allowing large drug molecules to pass into the epidermis between the cells of the stratum corneum.[12]

The stratum corneum's layers of keratinised cells limit the rate of absorption of constituents more than the thin respiratory epithelium. Where the stratum corneum is thinnest, for example on the face and scalp, constituents will penetrate the stratum corneum faster. Constituents are also thought to penetrate the skin via the sebum-lined hair follicles. The areas of skin with the greatest concentration of hair follicles are the forehead and the scalp, although most of the body has hair follicles. The palms of the hands have relatively thick epidermis and no hair follicles, which means that essential oils will preferentially be absorbed by the thinner body epidermis of the person being massaged, rather than the masseur.

Once they have penetrated the stratum corneum, or travelled down the hair follicles, constituents pass into the epidermis and dermis. Essential oil constituents are absorbed into the capillary circulation at the level of the dermis (about 5–6 mm below the surface of the skin). It is not known to what extent essential oil constituents penetrate to the adipose layer from a dermal application, though dermally applied oils do appear to have anti-inflammatory and analgesic effects on muscles and joints without having to enter the systemic circulation.

In a study on drug percutaneous penetration enhancement by sesquiterpenes, Cornwell and Barry (1994) found that the enhancement effects of the sesquiterpenol nerolidol lasted for about four days.[13] This implies that sesquiterpenes are not rapidly cleared from the skin layers and may be poorly absorbed into the circulatory system from a dermal application, in contrast to inhalation.

Percutaneous absorption enhancement
Due to their skin penetrating abilities, a number of different terpenoids are being examined as potential carriers for transdermal drugs. Table 5.2 offers examples of some research on essential oil constituents in this area. Nearly all of the components tested are monoterpenoids, with nerolidol being the only sesquiterpenol. The application of essential oils to skin near trans-dermal patches, such as oestrogen or nicotine patches, should be avoided. 1,8-cineole

Table 5.2   Summary of research on percutaneous absorption enhancement

Terpenoid	Drug	Increase in permeability of stratum corneum	Model	Author
limonene, eugenol and menthone	Tamoxifen	Increased	Excised pig epidermis	(Zhao & Singh, 1998)[14]
menthol	testosterone	8-fold increase	Excised human skin	(Kaplun-Frischoff & Touitou, 1997)[15]
terpinen-4-ol alpha-terpineol	hydrocortisone	3.9-fold increase 5-fold increase	Hairless mice	(Godwin & Michniak, 1999)[16]
1,8-cineole	5-fluorouracil	95-fold increase	Excised human skin	(Williams & Barry, 1991)[17]
nerolidol ethanol	5-fluorouracil	20-fold increase 13-fold increase	Exercised human skin	(Cornwell & Barry, 1994)[18]

was the most effective terpenoid at increasing penetration of 5-fluorouracil (a 95-fold increase).

Distribution

In pharmacokinetics, distribution can be defined as movement of drug substances between one location and another in the body (Bryant et al., 2003, p. 108). Figure 5.3 illustrates the pharmaco-kinetic pathway of drugs from absorption to excretion. (Note the reversible arrows indicating the constant movement of drugs as they progress from absorption to metabolism to excretion.)

Blood transport

As we have seen, essential oil constituents are lipophilic and insoluble in water. To be transported by the circulatory system, they have to be associated with lipids in the blood. Initially, they are likely to be associated with albumin or globulin lipoprotein molecules. Grass-man et al. (2001) found that terpenoid constituents of lemon oil were found in the low-density lipoprotein fraction of plasma after

Figure 5.3   Pharmacokinetic pathway of essential oils through different parts of the body from absorption to excretion

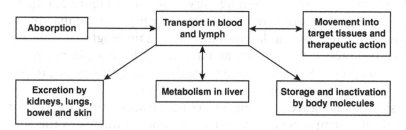

incubation with lemon oil.[19] Drugs that bind to plasma proteins are in a state of equilibrium between the bound and un-bound state. For most drugs, only the un-bound state is pharmacologically available, as the drug-protein complex is too large to diffuse through capillary walls.

Body fluid compartments

Figure 5.4 shows the distribution of drugs in the different types of body fluid, also known as 'compartments'. Note that drug molecules can only move from one compartment to another when they are freely dissolved in the fluid.[20] You can forcibly increase the amount of free drug in the plasma by upping or repeating the dose so that the

Figure 5.4   Distribution of drug molecules between body fluid compartments

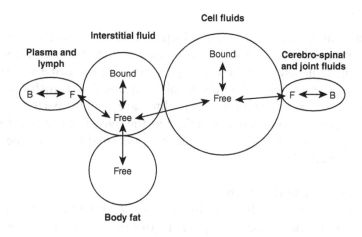

Source: Adapted from Rang & Dale, 1987, p. 71.

available plasma proteins become saturated with drug molecules and cannot bind any more drug molecules.

The plasma concentration–time curve shown in Figure 5.1 (page 117) presumes that drug molecules stay in only one compartment, and are eliminated by the kidneys after a certain length of time has elapsed. In practice this is rare, as most drugs have a plasma phase and a tissue phase. Some can have more than one tissue phase, if they have different affinities for different tissue types.

As shown in Figure 5.1, plasma concentration–time curves for essential oil constituents can have two or more gradients as the constituent is released back into the plasma compartment over time. Most pharmacokinetic analyses use a two-compartment model, with the plasma concentration in one compartment and the drug concentration of other tissues and fluids lumped together into the other compartment. Essential oil constituents have at least a bi-phasic distribution pattern, with alpha-pinene showing a tri-phasic pattern as demonstrated by Falk et al. (1990). They found that the initial distribution half-life of alpha-pinene was 4.8 minutes, with the second phase having a half-life of 38 minutes and a third half-life of 695 minutes (about 11.5 hours).[21] The existing pharmacokinetic analyses of essential oil constituents in the blood have not all considered the biphasic or triphasic distribution pattern noted by Falk et al., and thereby may be missing some important data, particularly concerning the length of time active constituents and metabolites of essential oil compounds remain in the body.

Metabolism

The main function of drug metabolism is to alter the water solubility of drug molecules so that they can be excreted in the urine although some drug molecules can be directly excreted in the urine so do not need to be metabolised. The chemical reactions that transform drug molecules into water-soluble metabolites are usually performed by enzymes. These enzymes occur mainly in the liver, but are also found in other parts of the body including the epidermal layers of the skin (Rawlins 1989).[22]

Phase I and Phase II reactions

There are two metabolic reactions, Phase I and Phase II. As you might expect, they usually happen in order. Phase I reactions are carried out mainly by the cytochrome P450 family of enzymes attached to microsomes within liver cells, but other oxidase enzymes and some reductase enzymes also are involved. Phase I metabolic changes include the breaking of C=C double bonds, the addition of –OH groups, de-esterification reactions and de-methylation reactions (important for the metabolism of phenyl methyl ethers). All terpenes, aldehydes, ketones, esters, ethers, oxides and coumarins have to undergo Phase I reactions before they are excreted.

Phase I metabolic reactions result in metabolites with hydroxyl or carboxyl groups attached to them. According to Rang and Dale (1987) some molecules, for example the ketonic steroid hormones, only become active after the Phase I reaction. The ketone carbonyl group has to be converted by reductases to an alcohol hydroxyl group.[23] This appears to be the mechanism by which terpenoid ketones are metabolised (Jäger et al., 2001).[24]

Limonene is a good example of an essential oil molecule that becomes more therapeutically active after Phase I metabolism. According to Hardcastle et al. (1999), limonene is metabolised to perillyl alcohol and perillyic acid, both of which have greater inhibitory effects than limonene in preventing the isoprenylation of proteins involved in cancer cell growth (see Chapter 8 for structures of limonene and perillyl alcohol).[25]

Phase II reactions result in the conjugation of Phase I metabolites with other molecules such as glucuronide and sulfate groups. The conjugated metabolites are nearly always less active than Phase I metabolites, and are excreted in the urine or the bile. If they are excreted in the bile, they can be reabsorbed into the blood stream which results in a longer time-course for the drug molecule. The most common conjugate molecule is glucuronide, attached by an enzyme known as UDP glucuronyl transferase (UDP, uridine diphosphate). The lungs and the kidneys can also carry out conjugation reactions for some drugs.

Molecules such as the phenols thymol and carvacrol and monoterpenoid alcohols can be conjugated directly without having to undergo Phase I reactions. This means they are more likely to be excreted rapidly. Sesquiterpenols may require Phase I reactions to be adequately solubilised.

Excretion

In pharmacokinetics, excretion of a drug is the clearance of the drug and its metabolites from the body.

Clearance

Clearance refers to the ability of an organ or the whole body to remove molecules from the blood stream. It can be divided into clearance due to formation of metabolites (CL_{met}), and clearance due to excretion (CL_{ex} or CL_r for renal excretion). Systemic clearance (elimination of the drug from the body) usually happens at a constant rate (given the stability of a person's physiology). The rate of clearance is usually defined as the volume of plasma from which a substance is completely removed or metabolised in a unit of time, for example, 90 litres per hour (L/h) or 200 mls per minute. The higher the CL constant, the faster the substance is removed or changed. This constant is multiplied by the plasma drug concentration (mg/L) to determine the drug's elimination rate (mg/h).

The liver is the main CL_{met} organ for lipid-soluble drugs like essential oils. A high hepatic clearance rate for a drug is when about two-thirds of the drug molecules are cleared from the blood in one pass through the liver. A low hepatic clearance rate is if less than 15 per cent of drug molecules are cleared in one pass through the liver. According to Bryant et al. (2003), a normal liver blood flow rate is 90 L/h.

Biliary excretion and enterohepatic reabsorption

Conjugated metabolites can be excreted in the bile. This then allows for their reabsorption via the small intestine, known as enterohepatic reabsorption. Tsai et al. (1999) demonstrated that aesculetine, a coumarin derivative (see Chapter 8, p. 187), is excreted in rat bile and then reabsorbed.[26] Kohlert et al. (2000) mention that menthol and linalool also undergo enterohepatic reabsorption.[27]

Excretion of constituents by the lungs

At the alveolar interface, constituents that are unbound to plasma proteins can diffuse into the lung cavity and be exhaled. An example of this is so-called 'garlic breath', because the garlic sulfide molecules easily diffuse across the alveolar membrane. Falk et al. (1990) noted that about 8 per cent of alpha-pinene inhaled from the air was exhaled unchanged.[28] With inhaled d-limonene, exhalation was less than 1 per cent (A. Falk-Filipsson et al., 1993).[29]

Human clearance rates of terpenoid compounds

Falk et al. (1990) also report a human systemic clearance rate for alpha-pinene of between 1.1–1.4 L/h/kg body weight, which would be equal to 66–84 L/h for a 60 kg male. As a human being contains approximately 5 litres of plasma, this means that the alpha-pinene would be totally cleared from the plasma in about 4.5 minutes. This is a very high clearance rate. Other constituents will probably show different clearance rates, depending on their affinity for non-plasma tissues. They also found that only 0.001 per cent of alpha-pinene was excreted unchanged in the urine.

Pharmacokinetic pathway of essential oils

Figure 5.5 summarises the pathway through the body essential oils take when applied to the skin. Note that if essential oils are given internally they are absorbed through the gut, and therefore taken to the liver for metabolism before they have a chance to pass around the body. If essential oils are absorbed through the respiratory system they enter the circulation through the lungs, and then follow the same path as those absorbed through the skin.

Determining the therapeutic dosage range of essential oils

In order to determine the therapeutic dosage range of a drug, it is important to know its pharmacokinetic profile from absorption to excretion for many different doses. Key parameters include:

- percentage of dose uptake or absorption;
- peak plasma concentrations and half-lives;
- type of metabolites formed (Phase I and Phase II);
- clearance rates; and
- excretion routes.

Kohlert et al. (2000) reviewed 15 papers on human pharmacokinetics of alpha-pinene, 1,8-cineole, camphor, delta-3-carene, limonene, borneol, menthol, thymol, trans-anethole and eugenol.[30] Unfortunately, they did not comment on therapeutic intentions and whether the dosages used in the experiments were effective.

It is also useful to determine the drug's specificity for its target site(s), and for other tissues. If, as for terpenoid constituents, a drug is retained for longer by adipose tissue than the target site it will be less potent, though not necessarily less efficacious, than a drug that

Figure 5.5   Pharmacokinetic route of essential oils through the body

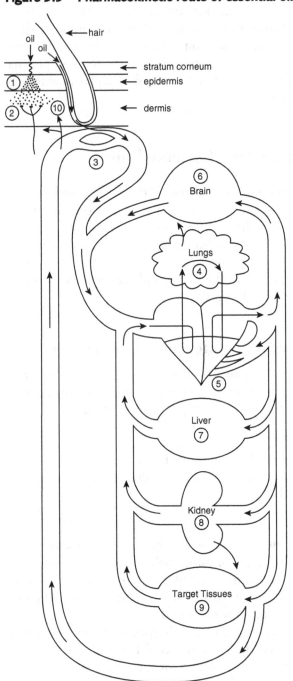

1. Metabolism of constituents by epidermal enzymes.
2. Interaction with nerve endings in dermis, and with inflammation.
3. Absorption into venous and lymph capillaries for return to heart.
4. Excretion of some constituents in lungs (minor).
5. First capillaries off the aorta are in the heart muscle i.e., the heart is the first internal organ to receive blood with essential oils.
6. Possible effects on the brain via neurotransmitters or other interactions.
7. Metabolism of constituents by liver enzymes — to make them more water soluble or easier to absorb into body. Storage in liver is possible.
8. Excretion of water soluble constituents and metabolites in urine. Lipophilic molecules returned to heart.
9. Storage or interaction of constituents in target tissues.
10. At the capillary level, essential oil constituents can diffuse into the skin. Some will be excreted from the skin, others will be stored in adipose tissue.

has high specificity for its target sites. Terpenoid molecules also seem to be less selective than some other pharmaceutical drugs, as they can affect several different target sites.

Once you know the pharmacokinetic profile, specificity and selectivity of the drug in question, you can experiment with varying dosages to observe the range of dosages that produce therapeutic effects. In order to determine a specific therapeutic effect, you need objective physiological measures of the target outcome. For example, to determine the therapeutic range of terpinen-4-ol for a candidal infection, you could take swabs of the infected area pre- and post-treatment, and compare the number of viable colony-forming units. To determine the therapeutic and safe range of a drug, dosages are increased until adverse effects start to appear, as these limit the usefulness of a drug. In the case of terpinen-4-ol, an adverse effect could be the appearance of irritation or sensitisation.

Dose–response curves
Dose–response curves can be plotted from these varying dose experiments as shown in Figure 5.6, using the terpinen-4-ol example. The curve flattens out when the maximal response is reached. Note that the number of colony-forming traits post-treatment is only an indirect measure of the therapeutic effects of the oil, and that the real measure of therapeutic effect is in symptom reduction. Maximal response *in vivo* is governed by the saturation of the available target receptors for the drug.

Steady-state pharmacokinetics
So far we have only considered the pharmacokinetics of single doses. Single doses are useful in situations such as local anaesthetic injections, where you want the effects to wear off. However most drugs require repeated dosages to achieve their effect. This is usually because the disease has thrown the body out of order, and it requires time and the continued presence of the drug to get back into balance.

Most pharmaceutical regimes are prescribed so that a given concentration range of the drug is present in the plasma at all times. This occurs when the rate of elimination is balanced by the rate of absorption of repeated doses. Figure 5.7 shows a hypothetical plasma concentration–time curve of repeated doses of the same drug.

Figure 5.6 A hypothetical drug concentration–response curve showing the dosage effects of terpinen-4-ol on numbers of colony-forming units of *Candida albicans* taken from swabs of the infected area (maximal response = 0 CFUs formed after treatment)

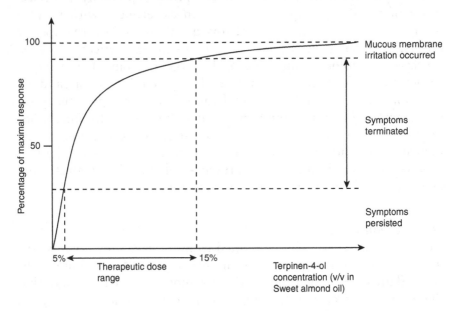

Figure 5.7 Plasma concentration–time curve after several repeated doses showing the steady-state pattern. Note how subsequent doses are given before complete elimination of the previous dose

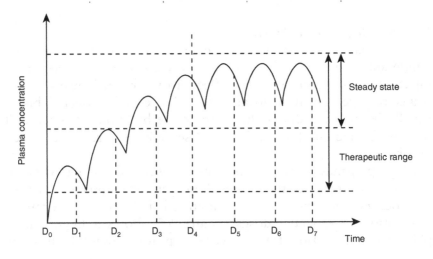

Therapeutic dose range of essential oils—is 3 per cent enough?

At the time of writing, there is relatively little research on the therapeutic range of most essential oils as defined above. The pharmacokinetics of most essential oil constituents are not well understood and, as the research by Falk et al. (1990) demonstrates, some of them display triphasic distribution patterns with long half-lives.

To compensate for this, most non-medical aromatherapy practice has followed the recommendation first formulated by Tisserand and Balacs (1995), which is based on extrapolations of rodent oral toxicity data for the most toxic essential oil studied, Boldo (*Peumus boldus*) leaf oil, and the least toxic essential oil studied, Valerian (*Valeriana officinalis*). The authors suggest that a lethal oral dose of Valerian for a 65 kg human would be about 1 kg of oil. Using their assumption that a maximum of 25 per cent of dermally applied doses penetrate the blood stream, they calculated that dermal application of 15 ml of a 3 per cent Valerian oil:vegetable oil solution should deliver a dose 2000 times less than the oral lethal dose, and therefore be safe.

This '3 per cent blend' (sometimes 2.5 per cent) has become a standard for aromatherapy dosage, with the understanding that you should reduce the dosage to 1 per cent for the elderly, seriously ill or child, and that you could increase the dosage to 5 per cent so long as you don't use more than 25 ml of the blend in any one treatment.[31] While adverse effects from this dosage range are exceedingly rare and are usually limited to idiopathic skin reactions, systematic research on whether 3 per cent is within or below the therapeutic dosage range in humans as shown in Figure 5.4 is yet to be carried out.

Another complicating factor in determining the therapeutic range of essential oils is that most aromatherapy treatments use more than one essential oil in the blend. This is often known as the 'art of blending', and borrows more from perfumery than from science. Certainly one of the positive effects of an aromatherapy treatment is the improvement of mood and well-being due to the pleasurable aroma, but in terms of pharmacology, the blending charts that I have seen leave much to be desired. As it is, we are already dealing with polypharmacy at a scale unimagined in mainstream pharmacy, with over 300 different constituents in each oil, most with unknown activities. We are also dealing with what I playfully term 'nano-polypharma-synergy', meaning the synergistic therapeutic effects of nanogram (10^{-9} g or 0.000000001 g) amounts of many different therapeutic substances, many of which show little if any activity in isolation.

Maximum tolerated dose (MTD)

Perhaps a more useful measure in determining the therapeutic dosage range of essential oils is the maximum tolerated dose (MTD) (calculated in g/kg of body weight/day). The MTD_{50} is the dose at which 50 out of 100 experimental animals experience the onset of adverse effects. According to Tisserand and Balacs (1995, Table 5.4, p. 48), the rat oral lethal dose LD_{50} for Valerian oil is 15 g/kg, whereas the oral MTD_{50} is 2.3 g/kg. This is almost seven times less than the lethal dose (LD).

The human MTD would be a very useful measure to establish for all essential oils (including chemotypes) and I suggest that any willing aromatherapist keep accurate records of all client data, dosages given and any adverse effects. Table 5.3 shows some examples of application methods and dosages used in non-medical aromatherapy, which are empirical measures of dosages that fall well below the MTD for most people.

Table 5.3 Application methods and dosages used in non-medical aromatherapy, assuming 1 ml of essential oil weighs 0.9 g, one drop weighs approximately 26 mg and an average human weighs 70 kg, and that absorption is complete (unlikely)

Application method	Volume of carrier	Percentage of essential oil/ number of drops	Length of contact with absorptive surface	Dose range (mg/kg)
Massage	10–25 ml vegetable oil	1–5%	5 mins–1.5 hours (skin, major; lungs, minor) can be up to 24 hours	1.2–16 mg/kg
Inhalation	500–1000 ml of hot water or in air	2–3 drops	5–15 mins (lungs)	0.7–1.1 mg/kg
Baths	450 litres of warm water	3–10 drops	15 mins–1 hour (!) (skin and lungs equal)	1.1–3.7 mg/kg
Dermal first (small burns and insect stings)	None–neat application	1–2 ml	5 mins (or longer if not washed off)	12.8–25.7 mg/kg

The massage, inhalation and bath dosages are designed to avoid skin/mucous membrane irritation as these are adverse effects (unless a counter-irritant effect is desired). For the inhalation and bath dosages (2 and 3), it is quite easy to discover for oneself that these dosage levels are the irritation limits for some essential oils (especially citrus oils or oils containing high proportions of thymol or carvacrol).

First aid application of undiluted lavender oil for burns and stings (dosage method 4) was developed as a response to emergency, when there was not time to blend the oils or measure the dose in drops, and for most people dermal application of 1–2 ml of undiluted lavender is well tolerated with no adverse effects.

Measurement of essential oil doses—are drops good enough?

For quantitation of pharmacological activity of essential oils, dose measurement techniques will have to become more precise than drops. Three problems associated with drop measurements are: different oils have different viscosity (thickness), meaning that the actual amount of oil in a drop of Sandalwood oil for example would be more than the amount of oil in a drop of Orange oil; the size of drops produced by the droppers and dropper inserts in bottles from different manufacturers varies; and different oils have different densities, so they do not weigh the same amount for a given volume. Svoboda et al. (2000) showed conclusively that 20 drops of essential oil (the commonly quoted calculation for 'how many drops in 1 ml') did not equal 1 ml for seven different commercially available dropper types and 11 different essential oils.[32] Twenty drops of most of the oils tested only came up to about 0.5 ml.

This is not to say that using drops to calculate dosage in general aromatherapy should be abandoned. In practice, using drops is the easiest way to create a blend for a full body massage. If a therapist consistently uses oils from the same supplier with bottles that deliver a constant drop size, they will be able to develop their own observations of efficacy based on numbers of drops from those bottles. Also, when making blends of essential oils for their overall perfume, drops are much easier to work with than millilitres.

Interactions of essential oils with pharmaceutical drugs

This area has not been well investigated, partly because the dosages of essential oils in aromatherapy are usually so small that the effects are negligible. However, that is not to say they don't exist. The theoretically possible oil–drug interactions listed by Tisserand and Balacs (1990, pp. 41–44, 66) are shown in Table 5.4.

Table 5.4 List of possible essential oil/drug interactions (adapted from Tisserand & Balacs, 1995)		
Drug type	Constituents	Possible interaction
Trans-dermal patches—nicotine, 5-fluorouracil (used in breast cancer chemotherapy) and the anti-inflammatory drugs Indomethacin and Prednisolone	limonene, camphor, 1,8-cineole, terpineol, nerolidol	Increases dermal penetration rates and absorption of drug
Warfarin and Heparin	eugenol, methyl salicylate, possibly coumarin and umbelliferone	Lengthens blood coagulation times, or prevents clotting
Pethidine, MAO (monoamine oxidase) inhibiting anti-depressants	myristicin, possibly other phenyl methyl ethers	Increases inhibition of MAO and therefore risk of respiratory failure
Paracetamol	trans-anethole, methyl chavicol (estragole), eugenol, methyl eugenol, safrole and apiole, d-pulegone, cin-namaldehyde—any molecule with a benzene ring	Increases risk of liver damage

Pharmacological targets of essential oil constituents

Now we have an understanding of how the body affects essential oils, and the mechanisms and processes at work, we are able to consider the pharmacodynamics of essential oils, that is, how essential oils affect the body. The actual target sites or target receptors are discussed, with a list of diseases or conditions that could benefit from binding of essential oil molecules to each type of receptor or site. These are summarised in Table 5.5 at the end of the chapter.

Interactions with lipophilic cell membranes

Every cell is enclosed by a phospholipid membrane, which keeps its contents distinct from the fluid between the cells. This membrane can be damaged by lipophilic solvents such as detergents, and by free radicals and oxidation. Essential oil constituents have activities in each of these areas, mainly due to their lipophilic nature.

Antimicrobial effects

The mechanism of anti-bacterial activity of monoterpenols and phenols is thought to lie in their damaging effects on the phospholipid membranes of the bacteria *Pseudomonas aeruginosa* and *Staphylococcus aureus* (Lambert et al., 2001).[33] When the membranes are damaged, the ion balance and pH levels in the cell are disrupted. Cells need to be able to control the ionic content and acidity of their intracellular fluids in order to maintain the correct conditions for cell life.

Antimicrobial effects of essential oils are cheap to research, and therefore there are a large number of reports on the antimicrobial effects of essential oils and their constituents. Research on terpinen-4-ol, the major monoterpenol in Tea-tree (*Melaleuca alternifolia*) oil, has shown it has both antibacterial and anticandidal properties (see Chapter 4). However, the effective dosage regime for *in vivo* infections has not been established for each essential oil.

Effects on human cell membranes

Hayes and Markovic (2002) found that the aldehyde citral, a major component of Lemon Myrtle (*Backhousia citriodora*) oil, is fairly toxic *in vitro* to human hepato-carcinoma (liver cancer) and normal skin fibroblast cell lines.[34] The dosage of Lemon Myrtle at which no adverse effects were observed over a 24-hour period was 0.5 mg/L in a liquid cell medium. They recommended that a lotion containing 1 per cent lemon myrtle oil would be safe enough and effective enough as an antibacterial lotion. Further research is needed to demonstrate what concentrations of other essential oil constituents may have similar membrane effects. It is unlikely that dermally applied essential oil constituents would achieve steady state plasma concentrations of 0.5 mg/L (equivalent to 500 ng/ml), but peak concentrations of 121 ng/ml have been observed from a dermal dosage equivalent to those used in aromatherapy.

Antiparasitic effects

Some essential oils like Wormseed (*Chenopodium anthelmenticum*) have been used in the past as treatment for intestinal parasites like worms and amoebae. However, the toxicity of the oil to humans (due to the oxide ascaridole) is a problem as Tisserand and Balacs (1995) report that the therapeutic dose is close to the toxic dose. The toxic effects of ascaridole include liver and kidney damage and brain oedema.[35] Most therapists would not use oils like Wormseed or Boldo (*Peumus boldo*), which also contains ascaridole. The antiparasitic mechanism may be as simple as dissolving the lipid membrane, in which case much less toxic oils could be used. The toxicity reactions appear similar to histaminic reactions (see below).

Solvent effects on cholesterol

Terpenoid constituents show good *in vitro* solvent properties for organic molecules. It may be that these solvent properties can be used to good effect in the body, if given in sufficient quantity, such as by injection, as the following example shows.

Dissolution of gallstones

Igimi et al. (1992) showed that limonene injected into the gall bladder of rats dissolved cholesterol-based gallstones.[36] Limonene may show similar effects on cholesterol-based atherosclerotic plaques in coronary artery disease, or in dementia, but it would most likely require dosages larger than those deliverable by dermal application or inhalation.

Anti-oxidant effects

Free radicals are molecules which have one less atom bonded to them than is required for molecular stability (that is, they have an incomplete outer electron shell), and will 'attack' other molecules which have C=C double bonds to try and get access to another electron to share and become stable. The phospholipids in membranes are susceptible to attack by free radicals (which are found in epoxy-resin fumes, car exhaust, fried foods and oils which are rancid or 'off').

Molecules such as alpha-tocopherol (also known as Vitamin E) are known as anti-oxidants because they interrupt the oxidation chain reaction started by the radicals. Phenolic compounds such as carvacrol, thymol, eugenol and the sesquiterpenoid phenolics from ginger oil, the gingerols, exhibit anti-oxidant properties in food, and possibly in living organisms (Aeschbach et al., 1994).[37]

Anticoagulant effects

Salicylates and coumarins are thought to have anticoagulant effects (see Chapter 4). More research is required to determine the safe and effective dosages.

Effects on excretion

As we have seen in the section on excretion, many essential oil metabolites pass out of the body with urine. The more rapidly a constituent is metabolised, the more rapidly it will be excreted. The pinenes, for example, have a high clearance rate and are found in high proportions in oils thought to be diuretics such as Cypress and Juniper.

Anti-oedema effects

Oedema is a build-up of fluid in extracellular space in tissues. It is often due to a malfunction of the lymphatic system, although there are many causes. Coumarins with hydroxyl groups may be useful for lymphedema management (see Chapter 4). The movement of tissue fluid may again be due to the flushing out of coumarin metabolites.

Bile stimulant effects

Limonene from the citrus oils is thought to be a bile stimulant, whether in the production of bile or in the release of bile from the gall bladder. This may be mainly due to the excretion of limonene glucuronides via the bile, rather than an actual increase in bile production per se.

Interaction with glutathione

As Tisserand and Balacs (1995, p. 61) note, glutathione is a molecule essential for the removal of free radicals and other toxic substances. It is found in most body cells, and concentrated in the liver. Some essential oil constituents taken orally (providing acute dosage levels compared to inhalation or dermal application) lower the liver's levels of glutathione: trans-anethole, cinnamaldehyde, methyl chavicol (estragole), eugenol, methyl eugenol, d-pulegone and safrole.[38] These constituents will compete with drugs which are also detoxified by glutathione and may result in an increased length of activity of the drug. Whether this effect would be significant for the chronic dosages of dermal application or inhalation is unknown.

Enzyme effects

Inhibition of nuclear transcription factors

Prevention of inflammation

Inflammation is a common symptom of many different disease processes. It is also the body's normal response to infection or injury, but it can cause unnecessary pain and discomfort, and damage surrounding healthy tissue. Control of inflammation decreases pain and suffering and can assist in the healing process. Inflammation is controlled by several different types of pro-inflammatory molecules, such as histamine, kinins and eicosanoid molecules (prostaglandins and leukotrienes). Anti-inflammatory drugs usually inhibit the enzymes that produce pro-inflammatory molecules, or antagonise their receptors.

In order to produce pro-inflammatory molecules, cells have to produce enzymes for their biosynthesis. These enzymes are encoded for by DNA, and are only transcribed when particular chemical messengers give the signal. Nuclear transcription factors like nuclear factor-kappa B (NF-κB) control the production of the prostaglandin synthase enzymes and other enzymes needed to create the pro-inflammatory molecules.

Several naturally occurring chemicals inhibit NF-κB. Yin et al. (1998) have determined that the anti-inflammatory action of aspirin (acetylsalicylic acid) is due in part to the inhibition of NF-κB.[39] The net effect is the inhibition of prostaglandin E2 formation from arachidonic acid. Lyss et al. (1997)[40] found that helenalin, a sesquiterpenoid lactone found in Arnica extracts, also inhibits the NF-κB transcription factors, but by a different mechanism than aspirin. The structure of helenalin, a sesquiterpenoid lactone (see Chapter 4), and that of aspirin and methyl salicylate (found in Wintergreen and Birch essential oils) are shown in Figure 5.8.

Figure 5.8   Structure of some anti-inflammatory molecules

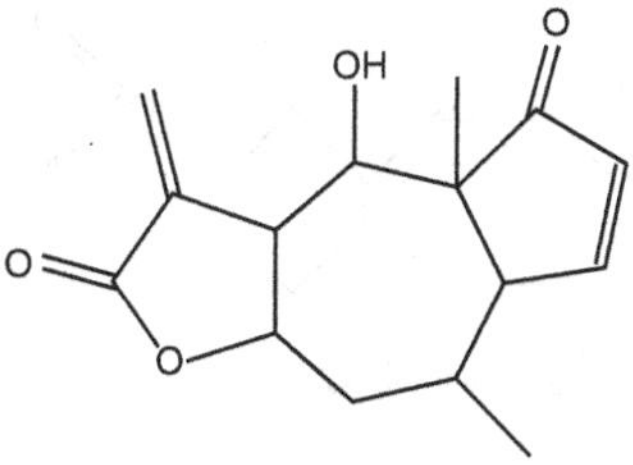

helenalin

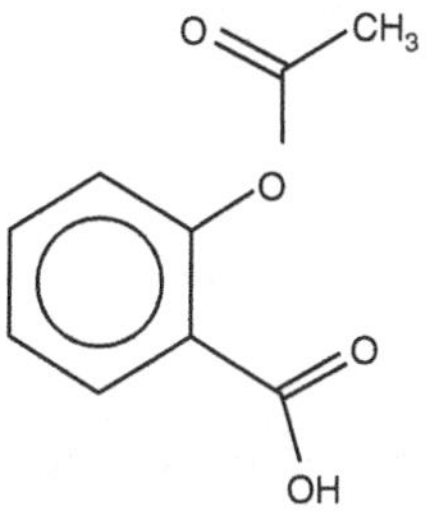

acetylsalicylic acid
(aspirin)

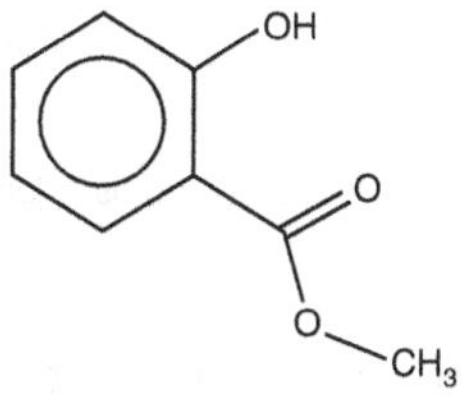

methyl salicylate
(Wintergreen and Birch oils)

Inhibition of cyclo-oxygenase and 5-lipoxygenase enzymes

Eicosanoid inflammatory molecules are made from arachidonic acid. Steroidal anti-inflammatory molecules like cortisone prevent the production of arachidonic acid, and non-steroidal anti-inflammatory drugs (NSAIDs) inhibit cyclo-oxygenase and 5-lipoxygenase enzymes.

Cyclo-oxygenases (COX-1 and COX-2) facilitate the formation of prostaglandins. There are three main types of prostaglandin: PGE1, PGE2 and PGE3. PGE2 is the pro-inflammatory prostaglandin, whereas PGE1 and PGE3 actually work in an anti-inflammatory manner. PGE2 intensifies the effects of histamine and the kinins, and summons phagocytes to the area. PGE2 is released by damaged cells in the dermal layer, and has the effect of prolonging and intensifying pain caused by stimulation of the pain receptors by kinin molecules (Tortora & Grabowski, 1993).[41] Figure 5.9 shows the structure of arachidonic acid and PGE2.

Figure 5.9   Eicosanoid molecules

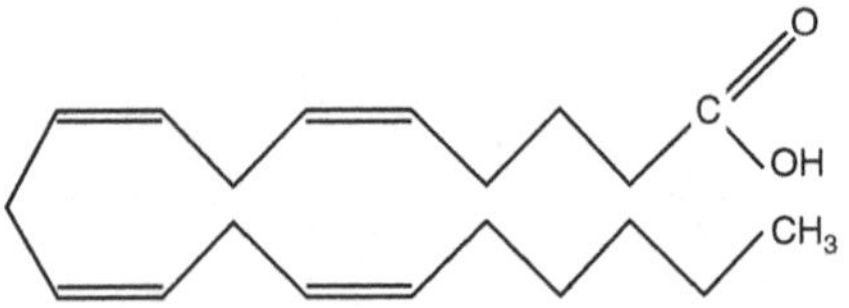

arachidonic acid

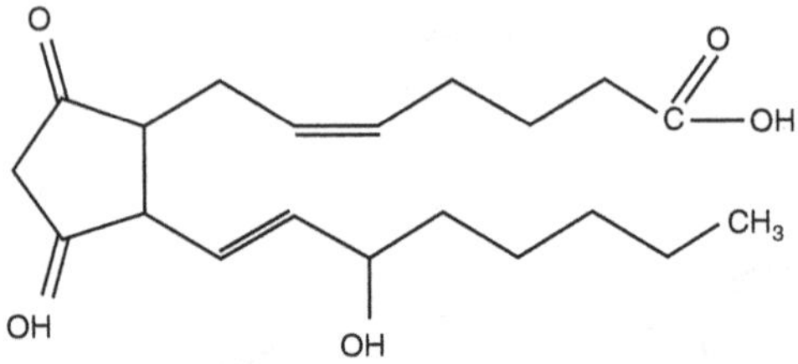

prostaglandin E2

5-lipoxygenase facilitates the formation of leukotrienes. Leukotriene B_4 is responsible for attracting white blood cells to the site of
inflammation, whereas leukotrienes C_4, D_4 and E_4 cause contraction
of smooth muscle and small airways, secretion of mucus, vasocon-
striction and make capillaries leaky, causing oedema (Rang & Dale,
1987).[42]

Anti-inflammatory effects
There is not much research on the effects of essential oil molecules on
eicosanoid molecules. Juergens et al. (1998) suggest that 1,8-cineole
may exhibit anti-inflammatory effects in bronchial asthma by
inhibiting synthesis of eicosanoid molecules from arachidonic acid in
monocytes.[43] Beuscher et al. (1998) found that a product known as
Gelomyrtol (containing alpha-pinene, menthol and 1,8-cineole)
also inhibits 5-lipoxygenase activity in basophils and neutrophils.[44]
It is likely that 1,8-cineole is again the active constituent.
    Safayhi et al. (2000) showed that triterpenoids from Frankin-
cense resin (Boswellic acids) inhibit 5-lipoxygenase enzymes *in
vitro*, but only in dosages greater than 10 µg/mL. Dosages between
1–10 µg/mL, in contrast, appeared to increase the production of
5-lipoxygenase.[45] Steam-distilled Frankincense essential oil is likely
to contain only minute amounts of Boswellic acids, so is unlikely to

have effects either way. However, liquid carbon dioxide extracts of Frankincense resin can be manipulated to have higher proportions of triterpenoids and may have potential in the treatment of inflammatory bowel disease and other inflammatory conditions.

Induction of adenylate cyclase

Antispasmodic and vasodilation effects

Adenylate cyclase is an enzyme that regulates the production of cyclic-AMP, an important 'second messenger' in the control of cellular function. Cyclic-AMP (c-AMP) upregulates enzymes known as kinases which activate other enzymes by phosphorylation. According to Rang & Dale (1987), c-AMP is usually associated with increasing the metabolic rate of cells, but it also has other effects on tissues: vasodilation; relaxation of intestinal smooth muscle, but contraction of respiratory smooth muscle; reduction of heart rate; inhibition of acetylcholine and adrenaline release and central neurons.[46]

Lis-Balchin et al. (2001) showed that Buchu oil, which is high in pulegone, appeared to exert its spasmolytic effect on guinea pig smooth muscle via upregulation of cyclic-AMP.[47] Qin et al. (1993) showed that oral administration of coumarins from *Cnidium monnieri* increased artificially lowered levels of prostaglandin PGE2 and cyclic-AMP plasma levels in rats treated with hydrocortisone.[48] Upregulation of c-AMP relies on an induction or enhancement of adenylate cyclase enzymes.

Induction or inhibition of cytochrome P450 enzymes

Some essential oil constituents increase the production of liver metabolic enzymes. According to Tisserand and Balacs (1995), geraniol, linalool, limonene and borneol all induce cytochrome P450 enzymes. The net effect of this is the acceleration of metabolism of these constituents. Pulegone on the other hand inhibits cytochrome P450 enzymes.[49] According to Crowell (1999), the anti-carcinogenic effects of limonene and other monoterpenes during the initiation phase of mammary carcinogenesis in rats are likely due to the induction of carcinogen-metabolising enzymes, resulting in carcinogen detoxification.[50]

Interactions with UDP-glucuronyl transferase

As this liver enzyme metabolises several types of essential oil constituent, it is likely that their presence in the body also induces its further production and activity. The enzyme is very important in detoxification reactions, particularly for constituents that have potentially toxic effects if not removed from the body, for example the neurotoxic ketone pulegone.

Fetotoxicity

Tisserand and Balacs (1995, p. 106) report three cases where mothers ingested camphorated oil while pregnant, and camphor was found in the bodies of the babies. In two of the cases, the full-term babies were still-born, whereas the third baby survived with no complications. Rabl et al. (1997) comment that camphor fetotoxicity is caused by foetal inability to metabolise camphor, as the liver lacks full development of UDP-glucuronyl transferase activity. There is no doubt that camphor crosses the placenta.[51] Other constituents that require conjugation with glucuronide may also pose similar risks to foetuses.

Inhibition of mevalonate pathway

Lowering blood cholesterol

Constituents like thymol and carvacrol lower blood cholesterol in animal models (see Chapter 4). This may be worth pursuing in human trials, although the dosages of these phenols required to achieve the effect may cause too much irritation on their way into the body, whether dermally or orally. The likely mechanism is interruption of the mevalonate pathway which controls cholesterol and steroid hormone synthesis. Cancer chemoprevention research on limonene and perillyl acid shows that both these monoterpenoids prevent the isoprenylation of enzymes in this pathway, but has not considered the possible reduction in cholesterol formation.

Fertility reduction

Essential oils used in aromatherapy do not appear to affect the development of the sex cells or gonads, though the aldehyde citral reduced female rat fertility when the subject was injected with the human equivalent of 25 ml of Lemongrass oil a day for 60 days (Tisserand & Balacs 1995, p. 104). Delgado (1993) reported the reproductive impairment of daughter rats born to mother rats who were fed myrcene in >5 g/kg doses while pregnant.[52]

It is likely that citral and myrcene inhibit some stage of the mevalonate pathway, preventing production of the steroid hormones required for normal development of the ovaries *in utero*. These are extremely high doses unlikely to be encountered in non-medical aromatherapy.

Inhibition of isoprenylation enzymes

Preventing tumour growth

Limonene and perillyl alcohol have shown anti-tumoral effects *in vitro* (see Chapter 3). Crowell (1999) reviewed the literature and noted that, in the post-initiation phase, monoterpenoid suppression of tumour activity may be due to the induction of apoptosis (cell death) and/or to the inhibition of the post-translational enzymatically controlled isoprenylation of cell growth-regulating proteins.[53] Research is in progress to determine how terpenoids compare in efficacy and toxicity with synthetic chemotherapy drugs.

Acetylcholinesterase inhibition

Alzheimer's disease

Acetylcholinesterase is the enzyme targeted in management of Alzheimer's disease. Its inhibition leads to elevated levels of the neurotransmitter acetylcholine, which is necessary for the formation of new memories.

While most essential oils have not yet been tested for acetyl-cholinesterase inhibition, Spanish Sage oil (*Salvia lavandulaefolia*) was tested by Perry et al. (2000) and showed *in vitro* activity similar to synthetic anti-cholinesterase drugs (an assay concentration about 0.03 μg/mL).[54] Perry et al. (2002) showed the oil to have *in vivo* activity in rats dosed with essential oil in sunflower oil for five days.[55] Although Rosemary and Sage (*Salvia officinalis*) essential oils are reputed to have memory-stimulating effects, they only exhibited mild anti-cholinesterase activity compared to Spanish Sage oil (Perry et al., 1996).

Other effects of acetylcholinesterase inhibiting drugs include lowered heart rate and blood pressure, stimulation of secretions in the salivary glands, respiratory passages and gut, increase in peristaltic motion and bronchoconstriction (Rang & Dale, 1987).[56]

Interactions with DNA and developmental processes

Any interactions of essential oil molecules with DNA are likely to cause cellular malfunctions and possible mutations that could also be carcinogenic. Novel plant substances are usually screened for this sort of toxicity in tests called mutagenicity tests. These test the ability of the substance to cause mutations in bacterial DNA (that is, is it a mutagen). If a substance is a mutagen, it is then usually examined further to see if it has adverse effects on developing embryonic cells. If the mutagen does cause deformities it is known as a teratogen. Teratogenicity is usually tested in rodents by breeding them with oral dosages of the teratogen.

Phototoxicity

As we saw in Chapter 4, the group of essential oil molecules known as psoralens, for example bergaptene (5-methoxypsoralen or 5-MOP) and bergamottine (found in cold-pressed Bergamot peel oils), can cross-link to DNA molecules in skin cells, causing cellular malfunction. This is a localised mutagenic effect, as it is limited to the site of application. However, psoralens have been used in the treatment of psoriasis, as they kill off the over-proliferation of cells that form the scaly skin characteristic of the disease (Fairlie et al., 1998).[57]

Hyperplasia

Engelstein et al. (1996) found that dermally applied citral, in synergy with high testosterone levels, caused prostatic hyperplasia in adolescent male rats.[58] It is not known whether this effect is found in humans, or whether the hyperplasia would lead to cancer in humans.

Developmental defects

According to Tisserand and Balacs (1995), sabinyl acetate found in Plectranthus oil (*Plectranthus fruticosus*) causes resorption of embryos and fetotoxicity in rats at oral dosages equivalent to human doses of 1 g/day. Plectranthus oil also causes birth defects in mice, and again this is thought to be due to the sabinyl acetate. The only essential oil containing sabinyl acetate that might be used in aromatherapy is *Salvia lavandulaefolia*, Spanish Sage, though the percentages vary between 0–24 per cent.[59] The biochemical mechanism of the defects is not yet established, but is likely to be due to interference with DNA.

Cell membrane ion channel effects

Calcium ion flow inhibition

Several essential oil constituents appear to inhibit the flow of calcium ions into cells. Calcium ion (Ca^{2+}) influx is controlled by special channel proteins in cell membranes. Voltage-sensitive Ca^{2+} channels open in response to depolarisation of the cell membrane caused by the incoming electrical impulse. These channels occur in neurons, heart muscle and smooth muscle. Receptor-operated Ca^{2+} channels are controlled by the binding of agonists to specific receptors.[60] To inhibit Ca^{2+} flow, essential oil molecules would have to either be antagonists for receptor-operated channels or interact in a physiological way with the membrane and prevent depolarisation.

Topical anaesthetic and analgesic effects

The nerve endings responsible for detecting pain, temperature and pressure are, for the most part, in the dermis. Menthol in Peppermint oil has a cooling effect on the skin at the site of application, which is sometimes described as warming. Reid et al. (2002) showed that menthol sensitises cold-sensitive neurons, thereby increasing the impression of cooling when, in fact, there is no change in temperature.[61] Galeotti et al. (2001) investigated the local anaesthetic activity of (+)- and (–)-menthol, and found that both compounds have a local anaesthetic activity, which they suggest is due to calcium ion blocking mechanisms.[62]

Eugenol from Clove oil has a definite temporary anaesthetic effect which has been used to good effect in dentistry (Seltzer, 1992).[63] Ghelardini et al. (2001) further studied the mechanisms of eugenol and other constituents and suggest that the effect is likely to be inhibition of Ca^{2+} ion channels in the sensory neurons (Ghelardini et al., 2001).[64] Monoterpenols geraniol, linalool and terpinen-4-ol all have mild analgesic effects on bites and stings, and quite probably have similar inhibitory Ca^{2+} ion flow effects.

Antispasmodic effects

Luft et al. (1999) showed that dietary farnesol (0.5 g/kg/day) in rats reduced rat hypertension for up to 48 hours.[65] Zygmunt et al. (1993) also found that cadinol inhibits Ca^{2+} ion flow in rat aorta cells, reducing smooth muscle contractions. However cadinol was 10 000-fold less potent than nimodipine, a synthetic calcium channel antagonist.[66] Calcium channel antagonists such as amlodipine and nimodipine are synthetic arterial relaxant drugs for lowering blood pressure (Rang & Dale, 1987).[67]

Peppermint oil has also been studied for its effects on intestinal smooth muscle. In an *in vitro* study, Hills and Aaronson (1991) showed that Peppermint oil applied to the isolated smooth muscle of guinea pig intestine prevented Ca^{2+} ion flow and thus contraction of the smooth muscle.[68] A review of the use of Peppermint oil for management of the symptoms of Irritable Bowel Syndrome suggests that Peppermint oil can offer some relief, but the quality of the trials was insufficient to be able to claim a reliable effect (Pittler & Ernst, 1998).[69]

Eugenol and anethole prevent isolated skeletal muscle contraction (Albuquerque et al., 1995).[70] However, they were applied to isolated muscle fibre *in vitro*, and may not have the same effects *in vivo*.

Effects on heart rate

Tachycardia is often a symptom of toxicity, for example camphor, and methyl salicylate (Tisserand & Balacs, 1995, pp. 51, 54), though the mechanism is not yet defined. They suggest that it may be due to calcium channel antagonism. Bradycardia (lowered heart rate) is also a possible sign of toxicity if accompanied with nausea and dizziness.

Receptor interactions

Histamine receptors

Histamine is produced by mast cells in the dermal layer. When dermal cells are damaged it is released, initiating the inflammatory response. Preventing histamine release prevents inflammation. There are two types of histamine receptors in humans: H_1 in bronchial and intestinal smooth muscle and H_2 in the stomach and heart. Binding of histamine to H_1 receptors causes calcium ion influx to the cells, smooth muscle contraction of bronchioles, as in asthma, and vasodilation of arteries. Large doses of histamine cause blood pressure to drop. Binding of histamine to H_2 receptors activates adenylate cyclase, which controls production of cyclic adenosine monophosphate (cAMP) and is an important second messenger linked to increased metabolic rate and smooth muscle contraction, causing a release of gastric acid in the stomach and increased heart rate (Rang & Dale, 1987).[71] Figure 5.10 shows the structure of histamine.

Figure 5.10   The structure of histamine

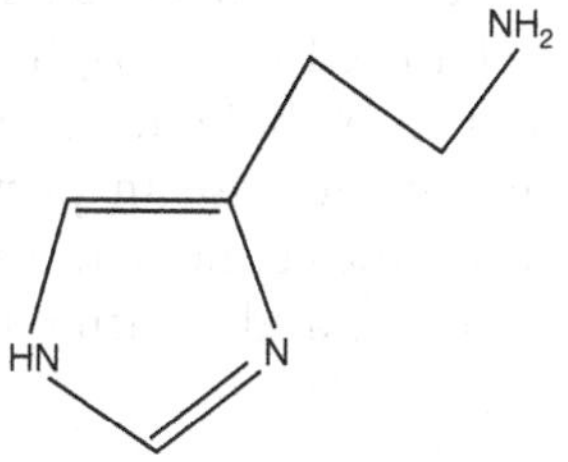

Antagonism of H_1 receptors

Antispasmodic effects

Essential oil constituents are likely to have effects on histamine H_1 receptors because of their antispasmodic action on smooth muscle tissue, but not a great deal of research has been completed on different constituents. Coelhodesouza et al. (1997) found that estragole and anethole blocked histaminic guinea pig ileum contractions (antagonism of H_1 receptors).[72] The calcium ion blocking effects of essential oils may also prevent the flow of calcium ions that occurs when histamine binds to H_1 receptors (see Calcium flow inhibition in Table 5.5).

Agonism of H_1 receptors

Allergic effects

Some dermal allergic responses to essential oil constituents may be due to histaminic receptor interactions, as shown by Santos et al. (2002), who examined the scratch-response of mice injected subcutaneously with 40 μL of 1,8-cineole. When they used an H_1 antagonist, the scratch-response caused by 1,8-cineole was greatly reduced.[73] This implies that 1,8-cineole irritates by mimicking the effects of histamine at H_1 receptors (that is, 1,8-cineole is a histamine agonist). More research is needed to ascertain whether this is a systemic effect, or whether it is specific to these experimental conditions (it is unlikely that you would ever get that amount of 1,8-cineole concentrated under the skin through any form of application used in aromatherapy).

The irritant effects of some essential oil constituents could be due to agonistic effects on histamine receptors. This would account for the counter-irritant effect, and for the fact that some people are allergic to perfumes and essential oils.

Counter-irritant or rubefacient effects

Constituents and oils that are thought of as irritants—phenols, monoterpenes, aldehydes—when used in the right dosages can create a mild dermal irritation, or the 'rubefacient' effect. Agonism of histamine H_2 receptors causes an increase in peripheral blood flow (vasodilation), thereby increasing the redness and warmth of an area. The effect is likely to be temporary, and treatment of the underlying causes is necessary.

Trigeminal nerve irritation

Trigeminal nerve endings respond to chemical irritation, pain, heat and cold. It is likely that agonism of H_1 receptors would stimulate the release of chemicals that sensitise the trigeminal receptors. The trigeminal nerve also has direct links to a part of the brain known as the reticular activating system (RAS), which is responsible for alertness and wakefulness.

Reviewing the use of odours in assessment of anosmia, Pinching (1977) found that people with lesions in the olfactory pathway causing anosmia can usually still perceive some odours due to trigeminal stimulation.[74] From the 1800s, the trigeminal stimulation of ammonia or 'smelling salts' has been used to restore consciousness to those who have fainted.

Aldehydes and phenols are known to have an irritant effect on mucous membranes, and it is likely they can therefore increase alertness just by stimulating the trigeminal nerve. Oils traditionally used in aromatherapy to increase alertness include Peppermint, Basil and Rosemary. Menthone, methyl chavicol and 1,8-cineole are probable irritants to the trigeminal nerve, which may explain the stimulating effects of these oils.

Oestrogen receptors

Several essential oils are reputed to have 'emmenagogic' (i.e., hasten the onset of menstrual flow) or 'oestrogenic' properties. This seems to imply that their constituents interact in some way with the reproductive hormonal cycle, either as agonists or antagonists for progesterone and/or oestrogen.

Howes et al. (2002) showed that citral, geraniol, nerol and trans-anethole all had oestrogenic activity when tested *in vitro* on yeast cells genetically modified to express human oestrogen receptors. However, the dosages required were quite high, and the researchers queried whether there was any clinical application in humans of these results.[75]

Perry et al. (2001) showed that Spanish Sage oil also had *in vitro* oestrogenic activity at dosages of 0.01 mg/ml, while geraniol showed milder oestrogenic activity.[76] Both Rose and Geranium oils are supposed to be helpful in regulating or balancing hormonally related conditions like PMS and menopause.

However, no randomised controlled trials have been done that conclusively indicate whether the oils do in fact make a clinically significant difference. Both oils contain similar proportions of the monoterpenols citronellol (about 20–30 per cent) and geraniol (about 15–20 per cent). Of note here is that the mevalonate pathway, which has many outcomes (one of which is the manufacture of steroid hormones), includes molecules known as geranyl geranyl pyro-phosphate. Perhaps the supplementation of extra geraniol increases the production of oestrogen as well as the molecule having oestrogenic activity itself. Figure 5.11 shows the structures of the main oestrogen molecule, beta-estradiol and progesterone, which also controls the reproductive cycle.

Figure 5.11 Structures of two female steroidal hormones

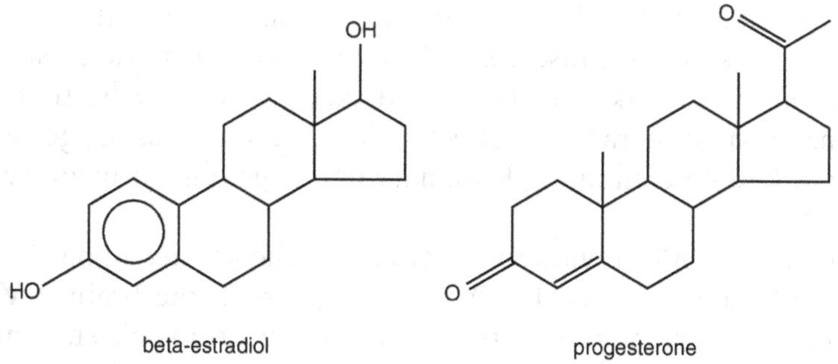

beta-estradiol   progesterone

Essential oils such as Clary Sage, which are supposed to be emmena-gogic, have not been proven to be so in any form of clinical trial that I know of. However, the structure of a compound found in Clary Sage known as sclareol looks very similar to the steroid molecules.

The concrete of Clary Sage (i.e. solvent extracted) contains 38–45 per cent sclareol, whereas the essential oil usually contains only a few traces of it (Schmaus & Kubeczka, 1985).[77] Figure 5.12 shows the structure of sclareol, which resembles the steroidal structure of the sex hormones. It remains to be tested whether sclareol has any oestrogenic effect and, if so, at what dosage.

Figure 5.12   Structure of the diterpenol, sclareol

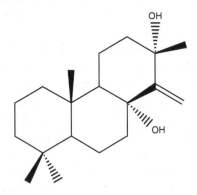

Neurotransmitter effects

GABA inhibition

Apart from the trigeminal effects, it is possible that essential oils can also affect neuronal activity within the brain. Given that a number of terpenoid ketones cause central nervous system toxicity such as convulsions when taken in large oral dosages, it may be that they could have some therapeutic effect at lower dosages as suggested by the use of ketone-containing Rosemary and Sage oils for memory and alertness.

Hold et al. (2000) reviewed the effects of the ketone alpha-thujone (found in Sage oil) on GABA A-type receptors in the brain.[78] When these receptors are bound by their normal neurotransmitter, gamma-aminobutyric acid (GABA), the binding has the effect of damping down the stimulation present in the brain. However, when alpha-thujone binds, it blocks the brain's ability to 'turn down the volume' and the excitatory effects of glutamate proceed unchecked, leading to over-stimulation, epileptic fits and, possibly, death.

Glutamate receptor antagonism

We have seen that some essential oil constituents (for example, monoterpenol linalool) have definite sedative effects on neuronal function. These sedative effects appear to be caused by antagonism of N-methyl-D-aspartate (NMDA) receptors (Brum et al., 2002).[79]

Nicotinic acetylcholine receptor agonism

Perry et al. (1996) showed that Spanish Sage has nicotinic acetylcholine receptor (AChr) agonist effects, which would contribute to alertness and the ability to form new memories.[80] It has obvious implications in the treatment of Alzheimer's disease and dementia. Other effects of nicotinic AChr agonism include increased secretion of adrenaline and stimulation of the 'flight or fight' aspect of the sympathetic nervous system (Rang & Dale, 1987).[81]

Psychopharmacology

Indirect effects of essential oils on mood and cognition via the olfactory pathway

The study of the effect of aroma on the body involves three levels of examination:

- the chemical interactions of essential oil molecules and olfactory neurons which lead to the perception of a smell;
- the links between the olfactory system and the limbic system which controls mood and memory; and
- the psychological effects of those interactions, or what we make the smells mean.

The olfactory system

Odorous substances are breathed in through the nose and dissolve in the olfactory mucus layer. Here they bind to G-protein receptors on the fine hair-like endings (cilia) of the olfactory nerves. When an odour molecule comes in contact with an olfactory G-protein, the protein registers its shape (or perhaps its bond energy vibrational pattern). If the molecule fits, the G-protein changes shape on the inside of the cell. This change of shape can trigger the generation of an electrical impulse or release a cascade of 'second messenger' molecules inside the cell (for example cyclic-AMP).

Impulses are generated and carried along the membrane of the olfactory neuron, and collected in a pattern of impulses in the olfactory bulb. These information patterns are then relayed from the olfactory bulb to various areas of the brain, including the limbic system, hypothalamus and the anterior part of the pituitary gland. You only consciously know that you have smelled something when the messages reach the cognitive areas in the neocortex of the brain.

The limbic system, mood, odour conditioning and memory

The limbic system is a collection of neuronal structures deep within the brain that are thought to control various subconscious processes. The amygdala is held responsible for fundamental mood states such as rage and fear, whereas the hippocampus is associated with laying down long-term memory patterns. The limbic system receives information from the olfactory bulb, even if the person is not conscious of an odour.

The pleasantness or unpleasantness (hedonic quality) of an odour is determined in the amygdala, and a reflex reaction occurs—either a facial expression, or turning the head towards or away from the source of the odour. Imagine a really disgusting smell being held up to your nose and notice how your nose and forehead instinctively wrinkle and you pull your head back. Very often our initial impressions of a person or place depend on these subconscious hedonic reactions to odour. Memories are often stored with olfactory components, such that a smell can immediately summon the full memory. Figure 5.13 shows the relationship between the limbic system and the other brain structures involved in olfaction.

Mood effects of odours

The hedonic quality of odours has definite effects on people's moods. Living in environments with odours that were perceived as unpleasant or bad caused people to report feelings of illness and annoyance (Steinheider, 1999).[82] Seeber et al. (2002) reported that chemically irritating odours increased people's self-reported levels of annoyance only when they were also perceived as foul smelling.[83]

By contrast, pleasant smells improved mood and productivity in the work-place. Workers in a pleasantly fragranced room chose more efficient strategies and set higher goals for themselves than those in a non-scented room (Baron, 1990).[84] Other research has examined the effect of essential oils on perceived mood. A major effect was that oils that were liked were more likely to cause feelings of relaxation, alertness, happiness and focus than oils that were disliked.

Odour conditioning

Another interesting possible psychological effect is odour conditioning. This is where a psychological or physiological challenge can be associated with an odour in test conditions, such that at a later date the odour alone causes a response as though the challenge were still being given.

Case studies from Birmingham, UK show that people with epilepsy can learn to associate the odour of essential oils like Ylang ylang with auto-hypnotic relaxation suggestions, and then use the inhaled odour of Ylang ylang to help avoid the occurrence of fits (Betts, 2002).[85] The work is still in early stages, and other oils have not been methodically tested to see whether there is a pharmacological effect as well as the conditioning effect.

Figure 5.13 The flow of information to different brain structures during olfaction

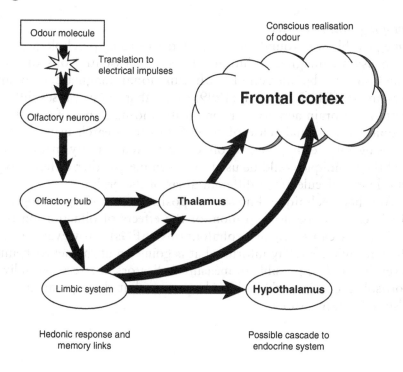

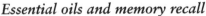

Essential oils and memory recall

Some essential oils are thought to stimulate memory formation and alertness. Moss et al. (2002) showed that the presence of Rosemary oil in the air of the testing cubicle enhanced subjects' short- and longer-term memory, but that it slightly impaired the speed of memory compared to controls.[86] The same study showed that the presence of Lavender oil caused a significant decrease in memory performance, and slowed reaction times in memory and attention tasks, compared to both Rosemary and the controls. It was not determined whether the longer-term memory effect persisted for more than a week.

Other studies have attempted to determine whether essential oils have reproducible effects on cognitive tasks, but have ended up asking more questions than they have answered. Part of the difficulty in assessing cognitive function is the variation in individual responses to odours, and the fact that not much is known about the dosage of different oils required to achieve a significant effect. Variability of constituents in any given oil type also makes it hard to compare trials that do not have an analysis of the constituents in the oils.

Imaging brain activity

One possible way round the complexities of using psychological tests is to use an imaging technique. Effects of essential oils on brain function can be measured with functional magnetic resonance imaging (fMRI). Levy et al. (1999) found that odour causes different patterns of brain activity in people with anosmia (loss of smell) than people with normal sense of smell.[87] Other research using fMRI is focusing on the mapping of olfactory-related activity in the brain, and the technique could be used to determine whether different types of odour molecule affect different areas of the brain.

Another technique known as positron emission tomography (PET) can also be used to monitor the effects of olfaction on brain activity, as can electroencephalograms (EEGs). However, none of these techniques really inform what is going on at the neurochemical level. It could be possible to measure indicators of stress like salivary cortisol levels, or perhaps blood pressure, but these parameters are also indirect measures.

Summary of health conditions and applicable pharmacodynamic effects of essential oil molecules

Table 5.5 on pages 156–7 lists different organs and body systems, and the health conditions associated with each that can be affected by essential oil molecules. Most effects seem to be beneficial, but some may be undesirable. It does seem that the polypharmaceutical nature of essential oils may prove to be 'just right' for treating the clusters of symptoms that often accompany complex disease states. It may be that they are best suited for the management of chronic diseases such as cardiovascular disease or arthritis because of their multiple modes of action on inflammatory processes, and the positive mood effects of the odours of different oils. Much could be said about the psychological benefits of an aromatherapy consultation and the relationship that develops between practitioner and client, but that is the topic of another book.

Further reading

- A practical, readable pharmacology text for health professionals: B. Bryant, K. Knights and E. Salerno (2003), *Pharmacology for Health Professionals*, Mosby/Elsevier (Australia) Pty Ltd, Marrickville, NSW.
- The Essential Oil Research Database, compiled by Bob Harris of Essential Oil Resource Consultants (EORC), is a useful source of scientific research concerning the biological activity of essential oils and their components. The Essential Oil Research Database is available in both a printed format that is updated annually and via annual subscription to an Internet version that is fully searchable and updated throughout the year. Subscribers also have access to a new database of GC and GC/MS analyses of essential oils compiled from scientific literature and actual commercial samples. http://www.essentialoilresource.com/index.htm [accessed 10 May 2003]
- A great resource for research into perfumery and the effects of odours on mood and cognition is S. Van Toller and G.H. Dodd (1991), *Perfumery: The Psychology and Biology of Fragrance*, Chapman & Hall, London. There is also a second volume, *Fragrance: The Psychology and Biology of Perfume* with more details. Both books are quite scientific.
- For an easy-to-read book summarising the effects of essential oils on the mind, emotions, soul and psyche see J. Lawless (1994), *Aromatherapy and the Mind*, Thorsons, London.
- For a fun read with all sorts of interesting information about the effects of odours on sexual response and reproductive behaviours in humans and other animals see M. Lake (1989), *Scents and Sensuality: The Essence of Excitement*, John Murray (Publishers), London.

Table 5.5 Health conditions and pharmacodynamic effects

System, organ or tissue	Condition/Effect	Non-specific effects: Cell membrane interactions	Lipophilic solvent effects	Antioxidant effects	Anticoagulant effects	Excretion effect	Interaction with glutathione	Enzyme effects: Inhibition of NF-κB	Inhibition of Cox and 5-lipoxygenase	Induction of adenylate cyclase	Antagonism of histaminic H₁ receptors	Agonism of histaminic H₁ receptors	Induction of cytochrome P450 enzymes	Interactions with UDP-glucuronyl transferase	Inhibition of mevalonate pathway	Inhibition of isoprenylation enzymes	Acetylcholinesterase inhibition	Interactions with DNA: Photoxicity	Hyperplasia	Ion channel effects: Calcium flow inhibition	Receptor interactions: Antagonism of H₁ receptors	Agonism of H₁ receptors	Oestrogen agonism	GABA inhibition	Antagonism of NMDA-glutamate receptor	Agonism of nicotinic acetylcholine receptors
Skin	Bites and stings							X	X		X									X						
	Burns							X	X		X									X						
	Peripheral oedema					X		X	X	X	X															
	Skin infections	X						X	X																	
	Psoriasis								X									X								
	Allergies	X		X				X	X			X					X			X		X↑				
Muscles	Arthritis							X	X			X								X		X				
	Cramps											X								X		X				
Respiratory	Respiratory infections	X						X	X								X									
	Sore throat	X						X	X																	
	Asthma							X	X		X						X			X	X	X↑				
	Hayfever							X	X		X	X										X↑				
Gastrointestinal	Inflammation of stomach and bowel lining							X	X	X	X															
	Intestinal spasm									X	X	X					X				X					
Circulatory	Poor circulation									X							X			X	X	X				
	High blood pressure									X										X		?				
	Deep vein thrombosis				X										X	X				X						
	High blood cholesterol		?	X											X	X										X

System, organ or tissue	Condition/Effect	Non-specific effects						Enzyme effects										Interactions with DNA		Ion channel effects	Receptor interactions					
		Cell membrane interactions	Lipophilic solvent effects	Antioxidant effects	Anticoagulant effects	Excretion effect	Interaction with glutathione	Inhibition of NF-κB	Inhibition of Cox and 5-lipoxygenase	Induction of adenylate cyclase	Antagonism of histaminic H_1 receptors	Agonism of histaminic H_1 receptors	Induction of cytochrome P450 enzymes	Interactions with UDP-glucuronyl transferase	Inhibition of mevalonate pathway	Inhibition of isoprenylation enzymes	Acetylcholinesterase inhibition	Phottoxicity	Hyperplasia	Calcium flow inhibition	Antagonism of H_1 receptors	Agonism of H_1 receptors	Oestrogen agonism	GABA inhibition	Antagonism of NMDA-glutamate receptor	Agonism of nicotinic acetylcholine receptors
Liver and gall bladder	Gall stones		X			X									X	X										
	Biliary insufficiency					X																				
	Hepatitis					?	X																			
Nervous system	Anxiety												X							X					X	X
	Insomnia																			X					X	X
	Drowsiness											X												X		
	Convulsions/epilepsy																							X		
	Headache							X	X		X									X	X					
	Alzheimer's disease		?	X	X			X	X	X	X		X				X						X		X	X
Dental	Toothache							X	X	X	X						X									
Reproductive	Uterine spasm			X																X						
	Breast and prostate cancer														X	X			X				X			
	Fertility reduction														X								X↑			
	Fetotoxicity					?							X↑	X↑												
	PMS & menopausal symptoms							X	X	X										X	X		X			
Kidney	Urinary retention					X				?											?					
	Genito-urinary infections	X				X		X	X																	

X = effect is listed
? = possible effect
X↑ = exacerbates condition

QUALITY CONTROL

Quality control of essential oils

Quality control of any therapeutic substance is sooner or later bound to become an issue. At present, the quality control procedures for essential oils are designed for the fragrance and flavour industry. Fortunately, these procedures also provide quality control information that is useful to aromatherapists. Quality control is important when you know the specific effects and hazards that you want to insure against, and current quality control methods are designed to ensure that the chemical composition of an essential oil fits within specified parameters. These parameters can come from publications like the medical pharmacopoeias, or from industry standards.

In 1993 I had the opportunity to work as quality control manager in a large essential oil importing company in Australia. What follows are insights I gained about quality control of essential oils during that time, along with those I have learned from other colleagues over the years.

In my discussions with aromatherapists, the main quality criteria they have is that an oil be a pure essential oil, from only one plant species, and from a single distillation. These principles are not necessarily adhered to in the production of essential oils for the fragrance and flavour industry, as the industry's main criteria are keeping costs low, maintaining reliability of supply and producing the right odour or taste.

Adulteration

Certain essential oils are scarce or expensive, so the temptation for producers is to add synthetic constituents, or constituents from other plant sources to 'stretch' the essential oil and thus increase profits.

Unfortunately, synthetic constituents cannot be guaranteed to be 100 per cent free of their starting materials. The usual guarantee for affordable synthetics is 95–99 per cent purity. Given the small amounts of essential oil used in dermal aromatherapy applications, it is unlikely that these trace starting materials would have a significant or acute effect, though chronic toxicity could perhaps be an issue.

Some essential oils are adulterated with other essential oils. The only problem with this is that it changes the percentages of the different constituents, and that some unwanted and unexpected constituents may be present. Most of the therapeutic properties of an essential oil have been determined by empirical and anecdotal evidence. If an oil is adulterated, however, you can no longer safely predict its therapeutic properties on the basis of your empirical knowledge, but have to rely on a chemical analysis of the oil. That is, unless the oil has always been adulterated.

How can we be sure we are buying good quality oils?

The cheapest way to assess quality is to train your own nose. You can do this by finding a reputable supplier and comparing the scents of their oils with oils that you know are synthetic, like cheap fragrance or room freshener oils. This will enable you, over time, to create a memory bank of what good essential oils should smell like. It does take time, however, and requires awareness of the different pre-extraction sources of variation as outlined in Chapter 2.

You should also be aware of the extraction process required for each essential oil, and the amounts of oil available from the plant. If, as with Rose and Jasmine oils, each ounce of essential oil is derived from 180 pounds of blossoms, you can be sure that these oils will be very expensive, unless they have been adulterated. This sort of information is available from perfumery and fragrance books, for examples J.W. Poucher's three-volume set *Perfumes, Cosmetics and Soaps* or journals such as *Perfumer and Flavorist*.[1]

Physical measures of essential oil quality

As well as assessing the odour quality of an essential oil, you can assess two physical measures using simple and inexpensive equipment. These are the 'refractive index' and 'specific gravity' or density of an essential oil, which can often appear on Material Safety Data Sheets supplied with essential oils.

The refractive index of an essential oil is a number that describes the way the essential oil bends light. A common example of light refraction is when you look at a straight stick in shallow water, and it appears to be bent at the point where the stick enters the water. This is because water has a different refractive index to air, so light rays passing into the water and out again will come out at a different angle, thus causing the illusion that the stick is bent.

A machine called a refractrometer can measure the refractive index of each oil, which can then be compared against a known standard. As each of the oil's constituents has its own refractive index, the refractive index of an essential oil is a composite, and depends on the percentage of different constituents within the oil. If the proportions of constituents with different refractive indices change, the oil's refractive index will change.

An oil's specific gravity or density is usually measured in grams per millilitre (g/ml) and, like the refractive index, depends on the oil's composition. Specific gravity is measured with a densitometer. Each essential oil has an acceptable range of densities at a certain temperature (usually measured at 20°C). The specific gravity of 1 millilitre of water at 20°C is 1 g/ml. At the same temperature, the specific gravity of most essential oils ranges between 0.78–0.90 g/ml, that is, they are less dense than water. This is why essential oils usually float on the surface of water. If the water is warmer than 20°C and the essential oils are colder, some will have a specific gravity that is heavier than that of water, and they will sink to the bottom. Oils that sometimes sink to the bottom despite temperature are Clove bud and Wintergreen oils.

An oil's refractive index or specific gravity cannot tell you precisely which constituents it contains, nor their precise proportions. To find this out, you need to use more sophisticated techniques: gas chromatography and mass spectrometry.

Gas chromatography

A more expensive (but more precise) way of finding out the constituents of a particular essential oil is to have a GC–MS analysis done. Gas chromatography (GC) is a technique that separates out the different essential oil constituents by their volatilities and polarities. A small sample of the oil is injected into a thin coated tube inside an oven which is slowly heated to about 250°C. The vaporised oil constituents are carried along the column in a flow of inert gas, usually helium. As they reach the end of the tube, they are identified by a special detector. A graph is printed out showing the different

Figure 6.1   Gas chromatograph of Tea-tree (*Melaleuca alternifolia*) oil

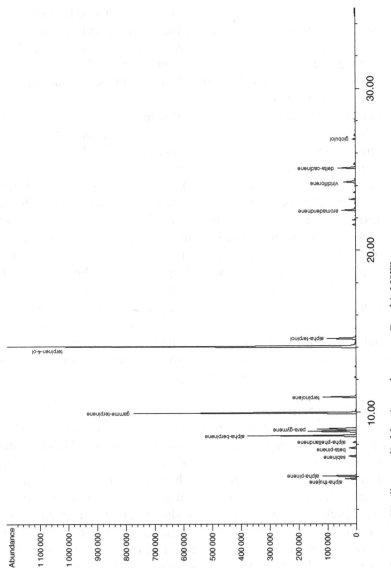

Source: Kindly supplied by Australessence, Coraki, NSW.

constituents as peaks separated by time, each peak representing a percentage of the total amount of oil tested.

The most useful type of gas chromatography is where the chirality of the constituents is taken into account. Natural oils generally contain constituents of homogenous chirality, whereas synthetic constituents will be a mixture of chiralities. (Chirality is connected with the arrangement of atoms in different isomers of the same compound, and the way in which the isomers refract light differently, i.e. their optical activity (see Chapter 7).)

Gas chromatography is usually combined with mass spectrometry. MS is the measurement of the molecular weights of a compound by first ionising them. The compounds, which can either gain or lose electrical charges, are then separated according to their mass-to-charge ratio, and detected. As each individual substance emerges from the GC it is fed into the MS, ionised and separated. The end spectrum is a series of peaks (a spectrum) where individual substances are separated in time and according to their mass-to-charge ratio. Figure 6.1 shows a GC–MS analysis of steam-distilled Tea-tree oil. Only some of the peaks have been identified, because they are the key components.

The x-axis is in minutes, calculated from the time the sample was injected into the GC machine. When the peaks are all on top of each other it is very difficult to distinguish the individual chemicals, even with mass spectrometry. It is possible to vary the peak separation by varying the heating rate of the GC oven. The more quickly it heats up, the more quickly the constituents come out but, if the analysis takes too long, chemical changes can alter the structure of the original molecules.

To find the constituents which are key odorants in a fragrance, the GC–MS system can have a second column which opens to the outside air at the same time as the peak is being drawn on the chart. This allows a perfumer or analyst to note which peaks carry which odour. This can be an even more powerful tool than the MS, as the human nose can detect some compounds at parts per billion.

Gas chromatography is less effective without the MS analysis, as to identify the peaks you first have to discover the retention times of each constituent in the oil. A way around this is to run a GC analysis of a standard sample, and then compare each subsequent batch to the standard. This technique is used for routine quality control by most essential oil companies. GC analysis can also easily determine whether there are any adulterants, though it requires a GC–MS analysis to find out what the adulterants are.

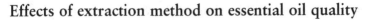

Effects of extraction method on essential oil quality

Other books cover the methods of extraction in greater detail, so I will give only a brief overview of each of the five main extraction methods used in industry, and comment on the ways in which each technique impacts on the chemical composition of the end product.

Steam distillation

Steam, or sometimes a combination of boiling water and steam, is passed through a container of plant material. The heat of the steam causes the oil glands to rupture and release their volatile oils. The temperature of boiling water is higher than the boiling point of most of the oils so they are converted into vapour, which can then be cooled in a condenser pipe and collected. The non-water soluble essential oil floats on the top of the re-condensed water in the collecting vessel, so can be siphoned off into separate bottles.

Steam distillation results in essential oils which have a high percentage of monoterpenoid and sesquiterpenoid compounds, and are volatile enough to be driven off at the temperature of steam (this depends on whether or not a pressurised system is being used). Temperatures in steam distillation do not usually exceed 100°C.

Steam distillation can affect the chemical composition of essential oils, as heat and water vapour can cause molecular rearrangements, hydrolysis of double bonds and, in general, will produce substances not originally found in the plant.

A common rearrangement reaction that occurs during distillation is the de-esterification of esters to their component alcohols and carboxylic acid. An example of this is during the distillation of Lavender oil, where the levels of linalool and linalyl acetate are key determinants of the overall fragrance of the oil. Linalool is sweeter, whereas the ester, linalyl acetate, is sharper and more refreshing in odour.

Depending on the length of distillation, you can alter the ratio of linalool to linalyl acetate and thus affect the end odour of the oil. A comparison was carried out between the oil produced from a hydro-distillation of lavender and a solvent extraction with pentane:diethyl ether (in a ratio of 7:3) (Schmaus & Kubeczka, 1985).[2] Table 6.1 shows the different percentages of linalool and linalyl acetate derived from the two extraction techniques.

Table 6.1   Ratio of linalool to linalyl acetate obtained with different extraction techniques

Compound	Hydrodistillation at pH 7.0	Solvent extraction
Linalool	41%	25%
Linalyl acetate	13%	43%

Source: Schmaus & Kubeczka (1985), p. 132.

A common occurrence in the fragrance industry is the addition of synthetic linalyl acetate to cheaper steam-distilled oil to make it smell more like the genuine plant fragrance, which is more closely emulated by the solvent extraction. The monoterpenols nerol and geraniol, and the various terpineol isomers also can be produced from the linalyl acetate during distillation. This can make the oil start to smell like Geranium or Tea-tree oil.

Distillers therefore try to minimise the length of time for which the constituents are actually held at those high temperatures. Time in terms of minutes would be the optimum for distillation, but obviously this depends on the quantity of plant material, and the industrial economics of such procedures. A general rule of thumb is, the smaller the still, the better the quality of the essential oil.

The formation of different chemical intermediates during the steam distillation process sometimes creates some unpleasant odours, known as 'still notes' (i.e. from the still), and an essential oil is often left to sit in drums for up to a year to allow the chemical reactions to stabilise and the still notes to mellow into the rest of the oil.

Cold pressing (for citrus oils only)

Cold pressing refers to the process of extracting oils from citrus peel (and is similar to that used for getting olive oil out of olives). The peels are put through a big masher and squeezed until they release their essential oil and watery juices. The essential oil is then skimmed off the top, and the waxes are usually allowed to settle out before the oil is packaged for sale.

Cold pressing obtains an oil which is close in odour to the original fruit peel as the process does not offer much chance for chemical reactions to take place. Unfortunately, the only plant materials that yield enough by this method to make it worthwhile are citrus peels.

It is possible to envisage cold pressing essential oil-bearing seeds such as fennel and black pepper, but the yield would probably be less than that from steam distillation, so the cost of the end product would probably be prohibitive.

Another reason for using cold pressing rather than steam distillation for citrus peel oils is that the acidic nature of any fruit material left on the peel is enough to catalyse rearrangement reactions which markedly alter the aroma of the oils. A good example is the comparison between cold pressed and steam-distilled Lime oil. The steam distilled lime has a quite distinct sherbert-like smell which is absent in the cold pressed oil.

Extraction with solvents

This method is more flexible than steam distillation, as you can tailor-make your solvent to ensure extraction of all the components you require, in particular the constituents with longer carbon chains such as the sesquiterpenoids, diterpenoids and coumarins. The types of solvents used are: liquid hydrocarbons such as hexane (C_6H_{14}); ethers such as diethyl ether; alcohols such as methanol and ethanol, and various mixes of these.

Chopped or ground plant material is placed in a container with the solvent and allowed to sit for a period of time (hours to days). The solvent is then removed by vacuum evaporation (which allows it to be removed at temperatures below the boiling point of the essential oil) or by chemical extraction. The remaining product is known as a concrete, and is often a complex combination of waxes, flavanoids and essential oils which is almost solid. The concrete is often further washed with ethanol to extract its terpenoid components and the ethanol evaporated, leaving behind a substance known as an absolute.

From a therapeutic viewpoint, it is impossible to remove every single molecule of solvent from the concrete or absolute, and this may pose problems of skin sensitisation or chronic toxicity. However, considering the amount of these sorts of substances that city-dwellers take in with every breath, I think we are unlikely to get many problems from remnant solvent molecules in absolutes, unless of course they are found to be carcinogenic, as was benzene, a previously used essential oil solvent.

Refrigerant and CO_2 extraction techniques

The answer to getting rid of the traces of solvent is to use a solvent which can be completely removed by evaporation at the end of the process. Two solvents have been developed that do this. One is the non-toxic, ozone-friendly refrigerant fluid 1,1,1,2-tetrafluoroethane (R134a) which has a normal boiling point of –26.2°C and a vapour pressure of 6.6 bar at 25°C, which means that at normal ambient temperatures and pressures, all of the solvent would return to its gaseous form. Dr Peter Wilde, of Wilde & Co. in England, has developed a sub-zero extraction technique using this solvent, which produces an extract containing mainly terpenoid constituents with traces of the pigments and waxes.

The extracts, known as florasols or phytols, are similar to the steam-distilled oils, but retain the delicate volatile components which get altered or lost in the hot conditions of steam distillation. As Dr Wilde says in his advertising material, the difference between florasols and essential oils is like the difference between freshly squeezed orange juice and marmalade!

The other non-toxic solvent which evaporates completely after the extraction process is liquid carbon dioxide (CO_2), a powerful non-polar solvent. There are two types of CO_2 extraction: subcritical and supercritical. In its subcritical state, liquid CO_2 behaves as a non-polar solvent. Under normal working conditions (temperature 0°–10°C, pressure 60–80 bar) it dissolves mainly non-polar and slightly polar components with molecular weights up to about 400. Aliphatic hydrocarbons up to fifteen carbon atoms (i.e. sesquiter-penoid), with hydroxyl, aldehyde, ketone and ester functional groups, are most readily dissolved (Moyler, 1988).[3]

It is also interesting to note that there are smaller percentages of monoterpenes in subcritical CO_2 extracts. This is probably because the process of steam distillation creates monoterpenes which do not normally exist in the plant. The example Moyler gives is that of Juniper essential oil and juniper subcritical CO_2 extract.[4] The essential oil contains 85 per cent monoterpene hydrocarbons and 2 per cent sesquiterpenes, whereas the CO_2 extract contains 69 per cent monoterpenes and about 10 per cent sesquiterpenes. From a flavourist or perfumer's perspective, monoterpenes are the packaging material for the oxygenated compounds which have the real flavours and aromas, so CO_2 extracts are prized as being about double the strength of essential oils.

Molecules of a molecular weight greater than 250 (i.e. larger than sesquiterpenes or sesquiterpenoid with more than one oxygen atom), carboxylic acids and some polar compounds with amino groups are only slightly soluble in liquid CO_2. Sugars, proteins, polyphenols, tannins, some waxes, inorganic salts, chlorophyll, carotenoids, citric and malic acids, and most alkaloids are not soluble in subcritical liquid CO_2.[5]

An easily perceived example of the difference between steam-distilled and subcritical CO_2 extraction is the comparison of the two types of extract of ginger (*Zingiber officinalis*). Several of the pungent principles of ginger known as shogaols and gingerols (see Figure 6.2) only occur in trace amounts in the essential oil, whereas in a CO_2 extract they are the major components. A smell or taste test will easily demonstrate the superior pungency and 'pep' of the CO_2 extract. The cool temperatures of subcritical CO_2 extraction also allow the collection of very volatile constituents which are lost in steam distillation. One of these constituents, hex-1-enal, gives the smell of freshly squeezed root ginger, and is found in the subcritical CO_2 extract but not in the essential oil.

Supercritical CO_2 extraction, on the other hand, does dissolve phenolic diterpenes, waxes, carotenoids and other plant constituents. A fluid is said to be supercritical when the temperature and pressure exceed the critical temperature and pressure for liquification, and yet the material remains a liquid. Supercritical CO_2 is denser than sub-critical liquid CO_2, and is a more powerful but less selective solvent.

Gerard et al. (1995) report that the anti-oxidant properties of supercritical CO_2 extracts of Rosemary (*Rosmarinus officinalis*) and Sage (*Salvia officinalis*) leaves (at 500 bar and 60°C) were more effective than those of the essential oil. This was possibly due to the presence of constituents which were not found in the essential oil.[6]

Figure 6.2　One of the isomers of gingerol

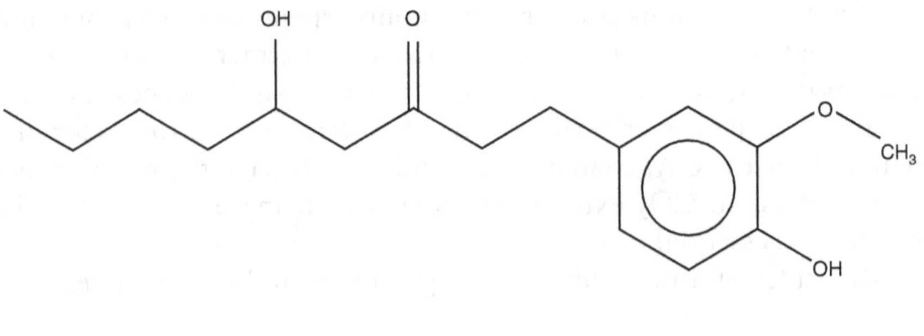

The inclusion of components such as waxes, carotenoids, flavanoids and alkaloids would make the therapeutic properties of supercritical CO_2 extracts more like herbal extracts or oleoresins than essential oils. I would caution against the replacement of essential oils with supercritical CO_2 extracts until pharmacological studies have been carried out which demonstrate their safety.

Enfleurage

This method is used for extracting constituents which are either known to degrade in the previous techniques, or which are present in such small quantities that every last little bit must be captured, for example, jasmine. The technique involves laying freshly picked blossoms onto a layer of fat (usually animal fat) on a sheet of glass, and leaving them until all the lipophilic materials (such as the terpenoid constituents) have been absorbed. This process is repeated daily for several days until the fat cannot absorb any more odour components, and requires many deft-handed workers to pick off the old blossoms without disturbing the fat layer. The resultant fragrant fat, known more elegantly as pomade, is then rinsed with ethanol to give an enfleurage absolute.

The disadvantages of this technique are that it can take up to two weeks to complete, it is highly labour intensive and only very small quantities can be properly processed at one time. Enfleurage absolutes are therefore very expensive. Precious floral extracts such as jasmine, narcissus and tuberose are still extracted by this process, but the solvent extraction methods are more common.

Infused oils and attars

Some essential oil companies supply infused oils and attars. Infused oils are vegetable oils that have had plant material from another plant left sitting in the vegetable oil for a period of time. They are a form of solvent extraction similar to enfleurage, except the vegetable oil is not separated off. Plants often made into infused oils are St John's Wort (Hypericum perforatum) and Calendula (Calendula officinalis), neither of which is available commercially as essential oils.

In India, essential oils are steam or hydro-distilled and the condensing vapours are allowed to condense into vats of Sandalwood (Santalum album) oil to form attars. The Sandalwood oil acts as a fixative for the more volatile essential oil constituents, so attars can be applied directly to the skin as perfumes or meditation blends.

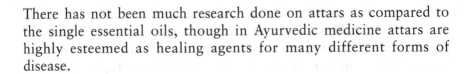

There has not been much research done on attars as compared to the single essential oils, though in Ayurvedic medicine attars are highly esteemed as healing agents for many different forms of disease.

Degradation of essential oils

Another issue in quality control is the degradation of essential oils. Some essential oil constituents, such as monoterpenes and mono-terpenoid aldehydes, readily combine with oxygen from the air, especially if there is any free energy around in the form of heat or light. Some will form resins (polyterpenes), others will be oxidised. The positioning of double bonds can be changed, open chains closed to form rings and the nature of functional groups changed, for example primary alcohols oxidising to form aldehydes. These processes will alter the therapeutic properties of the essential oil constituents, so it is important to store the essential oils in conditions which:

- minimise contact with the air. Small bottles with narrow necks and valve-like stoppers, or dropper bottles are best.
- minimise contact with free energy. Dark-coloured glass bottles and refrigeration prolong the shelf-life of the oils. Citrus and other mainly monoterpene oils must be refrigerated if being kept for longer than a year.

Another precaution is to not keep essential oils for more than about three years, as they do start to rearrange and degrade. However, no definitive studies have been done on the length of time for significant degradation of constituents, or on the effects of such degradation on the therapeutic qualities of the oils.

Oxidation of some oils may enhance their therapeutic properties. For example, in the case of Patchouli oil, the older the oil is the more efficacious it is considered to be, possibly because the aroma is richer and more rounded but probably because the oxidation of the molecules has made them more potent therapeutically. Further research could demonstrate this conclusively.

7

ISOMERS AND NAMING

This chapter extends our investigation into the structures and naming of essential oil compounds. One concept we have so far touched on only briefly in looking at molecular structure is that of isomers. When two molecules have the same formulae but different structures, they are known as isomers. For example, linalool has the formula $C_{10}H_{17}OH$, which is the same as the formula for geraniol (see Monoterpenols, Chapter 4), but the two molecules have a different arrangement of atoms and smell different. They also have different antibacterial properties, with geraniol usually being more powerful.

The differences between isomers can be more subtle than this, perhaps relating to the different positioning of a double bond. An example is alpha-pinene and beta-pinene found in Pine oil (see Common monoterpenes, Chapter 8). The overall three-dimensional shape is the same, the Greek prefixes *alpha-* and *beta-* indicating that there is only a minor difference between the two molecules. The prefixes are sometimes written as the Greek letters (alpha = α; beta = β). You can have many different isomeric structures for the same chemical formula. Types of isomers with importance for essential oil chemistry are geometric isomers and enantiomers.

Whether a molecule has enantiomeric forms or not depends on whether it contains an asymetric or chiral carbon atom.

Geometric isomers

Geometric isomers are associated with carbon double bonds. When you have a carbon double bond, there are several ways that the atoms adjoining the carbons can be arranged. If the longer chains come off the same side of the double bond, the molecule is called a cis-isomer. If they come off opposite sides, the molecule is called a trans-isomer. The simplified example in Figure 7.1 demonstrates cis-isomers and trans-isomers using the letter A to denote shorter chains, the letter B to denote longer chains. Another way of denoting cis- and trans- is (Z)- and (E)-, which stand for the German words *zusammen*, 'together', and *entgegen*, 'opposite'. Cis-isomers and trans-isomers have different chemical and therapeutic properties. From a quality control perspective, different essential oils have specific ratios of cis- and trans-isomers.

Figure 7.1   Cis- and trans-isomers

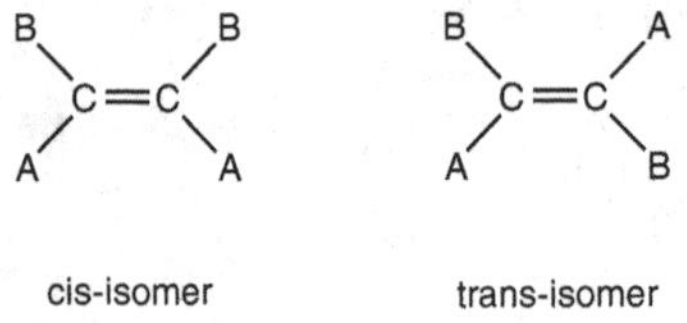

cis-isomer   trans-isomer

Enantiomers

If two molecules are three-dimensional mirror images of each other, and cannot be superimposed on each other, they are known as enantiomers. This is a bit difficult to visualise without three-dimensional models, but try anyway. An example of a pair of enantiomers is the carvone molecules shown in Figure 7.2. Note how the top parts of the molecules are more or less in the plane of the page due to the double bond; the 'tail' is where the variation comes in, preventing the superimposing of the molecules.

Chirality

Whether a molecule will exist in two enantiomeric forms can be predicted by determining the chirality of the molecule. This is done by looking at the groups bonded to the carbon atoms in the molecule. Carbon atoms with four different groups attached are known as chiral carbons. In the molecules of carvone in Figure 7.2, the chiral carbon atom is the one marked with an asterisk*.

Enantiomers often have different odours. The (+)-carvone shown in Figure 7.2 (Boelens et al., 1993) is the main constituent of

Figure 7.2 Enantiomers of carvone, showing chiral carbon atoms as *

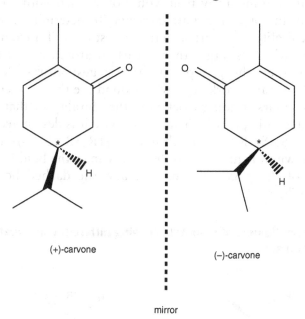

(+)-carvone

(–)-carvone

mirror

Caraway seed (*Carum carvi*) oil, whereas (–)-carvone is the main constituent of Spearmint (*Mentha spicata*) oil.[1]

Living organisms seem to preferentially produce or utilise only one enantiomer of pairs of enantiomeric molecules. The three-dimensional configuration of enzymes and the molecules they catalyse depends on this selective production of enantiomers. This poses problems in drug design, because laboratory reactions usually produce a mixture of the two enantiomers, known as a 'racemic' mixture. Therapeutically, a racemic mixture will be less effective than one with just the active enantiomer.

Enantiomers of essential oil constituents have been analysed for their different odours and tastes, but have not yet been systematically analysed for their therapeutic effects. I expect that we will see many instances of enantiomers having different therapeutic activities.

When a molecule has two chiral carbon atoms (where there are four different groups coming off each bond), you will sometimes see the use of R- and S- to differentiate them. These letters stand for *rectus*, 'right' and *sinister*, 'left'. This allows for comparison between two enantiomers, and the exact representation of their three-dimensional structure.

If you take the smallest group attached to the chiral carbon atom, and have that pointing away from you, you are left with an almost planar view of the other three attachments. To determine whether it has R- or S- chirality, you start with the largest atom (for example, an oxygen atom, which is larger than a carbon atom), and if not the largest atom, then the group with the largest atoms closest to the chiral carbon, and work your way around the three attachments.

If you find yourself going clockwise, the chirality of that carbon atom is designated R-; if you go anti-clockwise it is designated S-. An example is (1S,3S,4R)-(+)-menthol and (1R,3R,4S)-(–)-menthol, which are shown in Figure 7.3. In these drawings, the bold bonds are coming out of the page towards you and the dashed bonds are receding into the page.

Figure 7.3　Enantiomers of menthol, showing different configurations for chiral carbon atoms

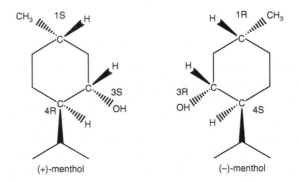

(+)-menthol　　　　(–)-menthol

Optical activity of enantiomers

An easy way to detect whether a pure liquid contains chiral carbon atoms is to test its optical activity in polarised light. Liquids containing molecules with chiral cabon atoms will rotate polarised light off-course when it passes through a tube of the liquid. Enantiomers are also called optical isomers.

Optical enantiomers will rotate polarised light in a particular direction, according to whether they are 'dextrorotatory' or 'levorotatory'. A pure solution of a dextrorotatory or (+)-enantiomer will rotate the light in a clockwise or right-turning fashion. A pure solution of the levorotatory or (–)-enantiomer will rotate the polarised light in an anticlockwise or left-turning fashion. If you have an equal mixture of enantiomers present (i.e. a racemic mixture) no optical rotation will occur, as the light is turned equally by the mirror image molecules. The molecules of carvone in Figure 7.2 are optically active enantiomers.

Most essential oils display some optical activity due to the different combinations of enantiomeric molecules they contain. Before GC–MS was developed, the optical rotation of an essential oil was used as an indicator of quality, as it reflects the oil's chemical composition. Even now, optical rotation is used as a quick routine test for essential oil quality. As mentioned in Chapter 6, there are now GC machines which can determine the chirality of different constituents, thereby allowing a greater fine-tuning of the quality assessment.

Systematic chemical nomenclature

So far we have been referring to essential oil molecules by their common names. This is useful for quick determination of which molecules we are dealing with, and probably is sufficient for aromatherapists using the oils on a daily basis. However, some scientific papers refer to molecules by their systematic chemical names, so I will give a brief example of some names and what they mean from a structural point of view.

Figure 7.4   Alpha-terpinene, showing numbering of carbon atoms and bonds

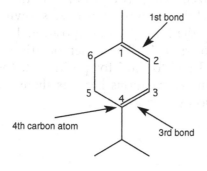

alpha-terpinene

Figure 7.5 A molecule of para-cymene or 1-methyl-4-isopropyl benzene, showing positions of groups attached to the ring

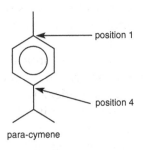

position 1

position 4

para-cymene

Alpha-terpinene (see Figure 7.4) is an example of the common name of a terpenoid molecule. Its systematic chemical name is 1-methyl-4-isopropyl-1,3-cyclohexadiene. In the systematic name, the last word usually refers to the bulkiest part of the molecule onto which all the other bits are attached. Here this is cyclohexadiene. 'Cyclo-' means there is a closed ring, 'hexa-' means there are six carbon atoms in the ring, and '-di-ene' means there are two double bonds in the ring. These double bonds are at the locations indicated by the numbers 1,3- preceding cyclo-. The numbers 1- and 4- indicate the location of the methyl and isopropyl groups on carbon atoms in the cyclohexadiene ring. The aim of the numbering is to keep the numbers as low as possible, so carbon number 1 in the ring is at the top where the methyl group is attached.

Another term, usually found in relation to molecules which have benzene rings, is 'para-', or sometimes just 'p-'. This refers to the position in relationship to each other of two functional groups attached to the benzene ring. An example is the molecule para-cymene, or 1-methyl-4-isopropyl benzene, shown in Figure 7.5. A chemist would say the methyl group (position 1) is 'para-' to the isopropyl group (position 4). The other positions are known as 'meta-' for position 3 and 'ortho-' for position 2. I will not go into other scientific naming conventions here, as there are other sources dedicated to the subject (see below).

Further reading

- An excellent source of information on the chirality of compounds and its effect on their odour is http://www.leftingwell.com/chirality/chirality.htm.

- G.J. Leigh, H.A. Farne and W.V. Metamomski (eds) (1998), *Principles of Chemical Nomenclature: A Guide to IUPAC Recommendations*, Blackwell Science, Malden, Massachusetts.

8

MOLECULAR STRUCTURES

In attempting to gather the structures for the most common essential oil constituents together as a reference I have used a variety of sources, including websites for chemical supply companies.[1] For ease of reference I have grouped constituents by their structure and functional groups.

Figure 8.1   Common monoterpenes

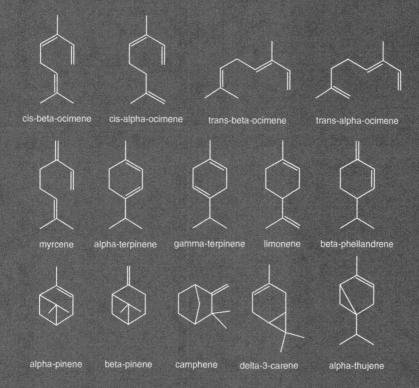

Figure 8.1 *continued*

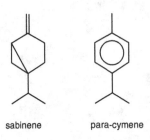

sabinene   para-cymene

Figure 8.2 Common sesquiterpenes

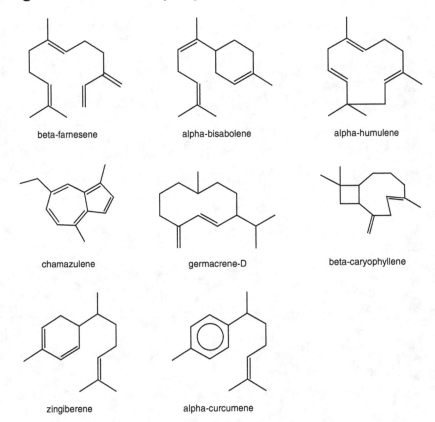

beta-farnesene   alpha-bisabolene   alpha-humulene

chamazulene   germacrene-D   beta-caryophyllene

zingiberene   alpha-curcumene

Figure 8.3   Common monoterpenols

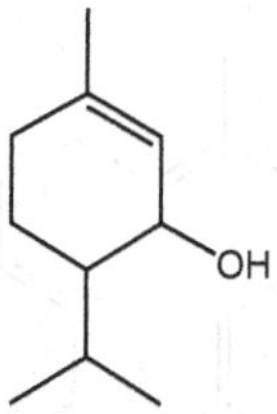

piperitol	menthol	alpha-fenchol	borneol
citronellol	nerol	geraniol	linalool
alpha-terpineol	beta-terpineol	beta-terpineol	terpinen-4-ol
perillyl alcohol	lavandulol	thujanol	

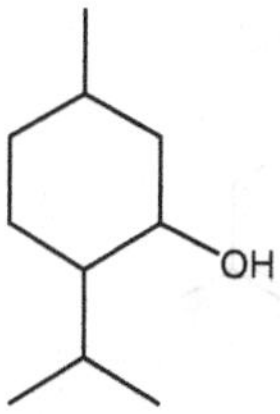

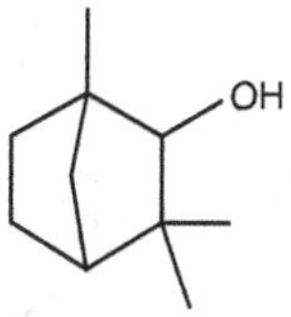

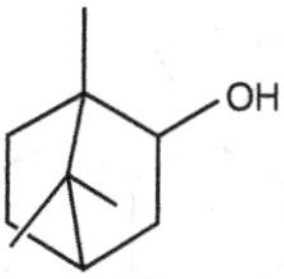

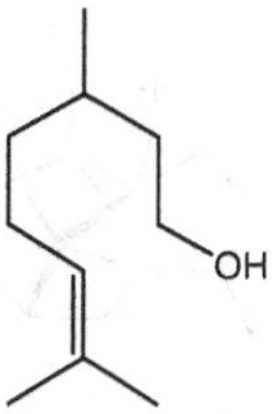

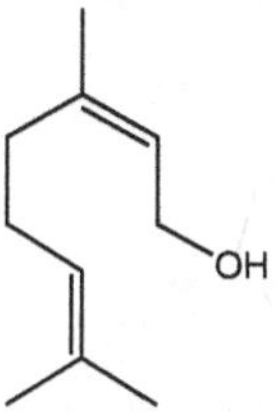

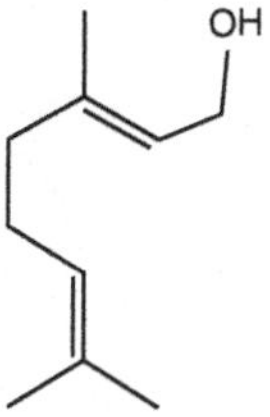

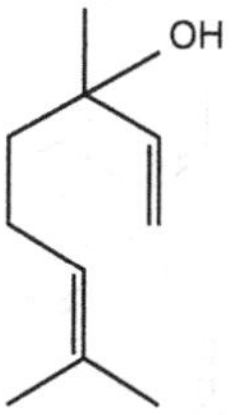

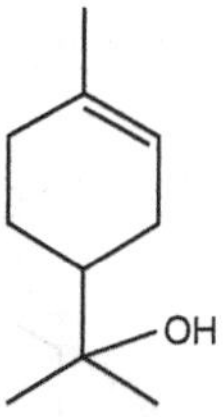

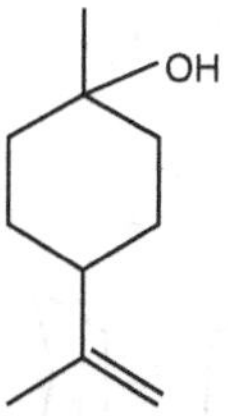

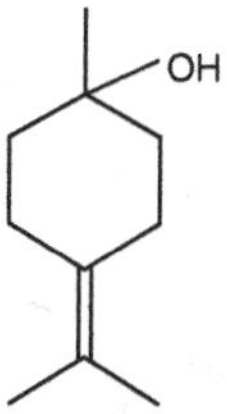

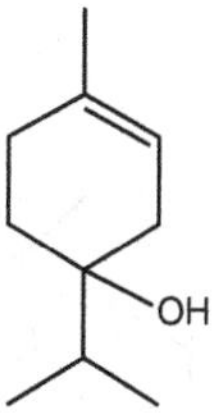

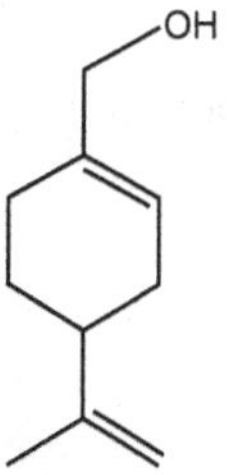

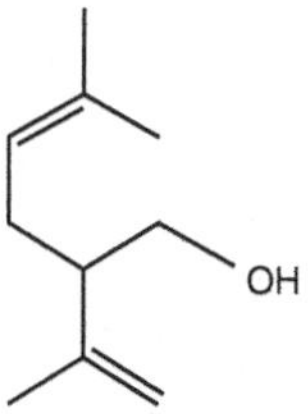

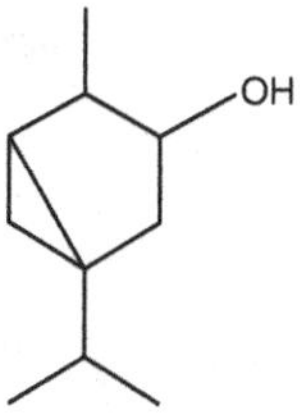

Figure 8.4   Common sesquiterpenols

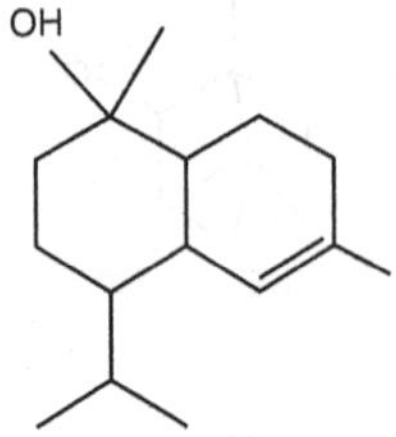

alpha-cadinol

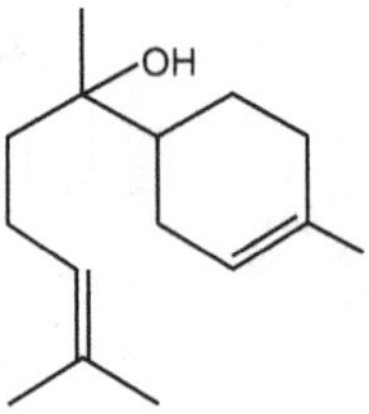

alpha-bisabolol

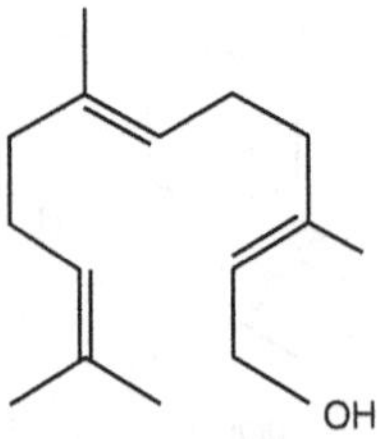

beta-farnesol

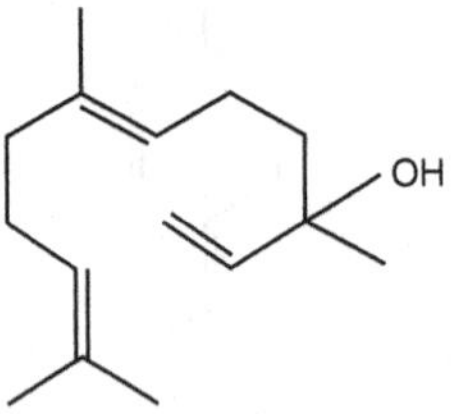

nerolidol

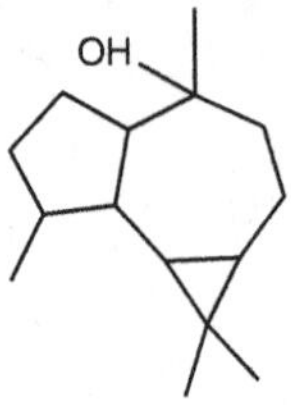

viridiflorol

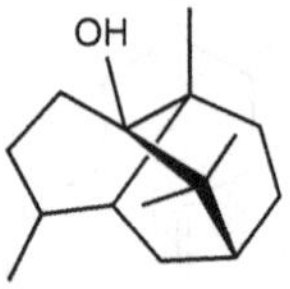

patchouli alcohol

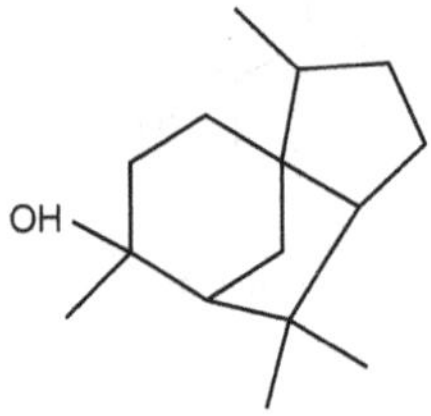

cedrol

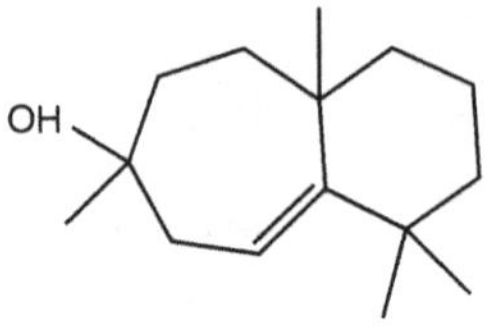

widdrol

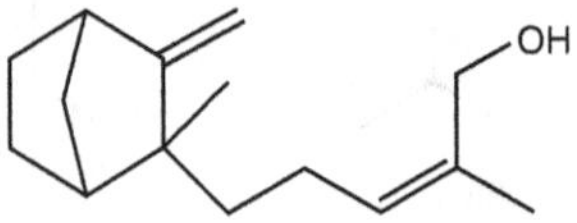

beta-santalol

Figure 8.5   Common phenols

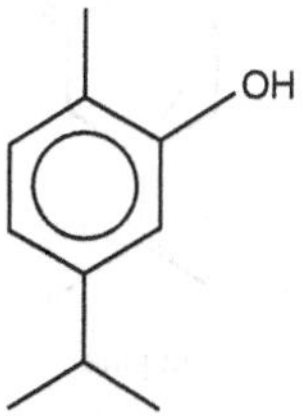

carvacrol	thymol	chavicol	eugenol

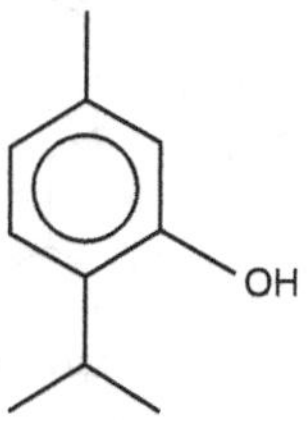

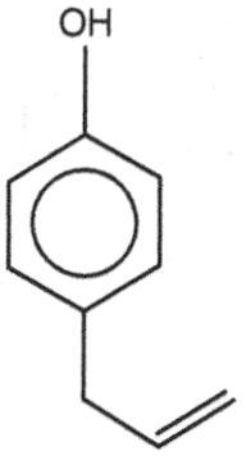

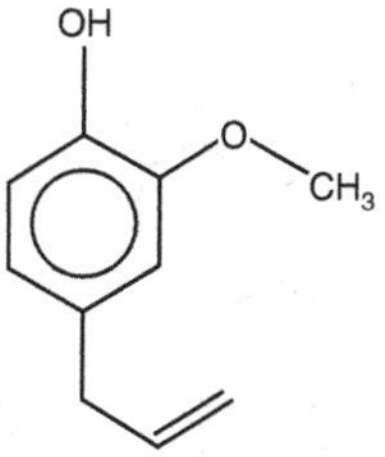

Figure 8.6   Common aldehydes

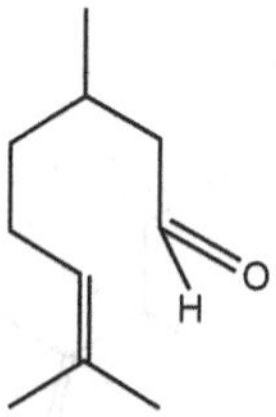

citronellal	neral	geranial	myrtenal

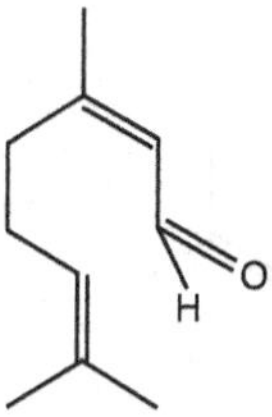

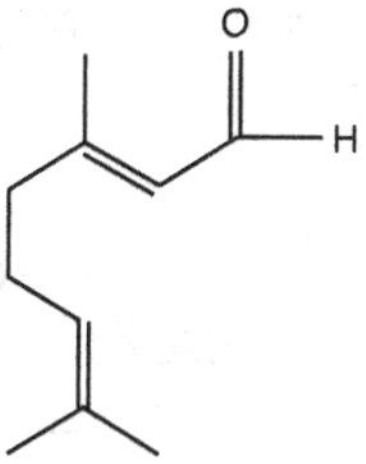

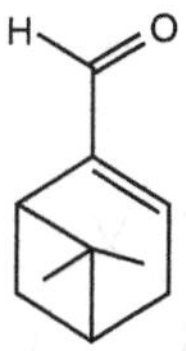

2-hexenal	decanal	cinnamaldehyde

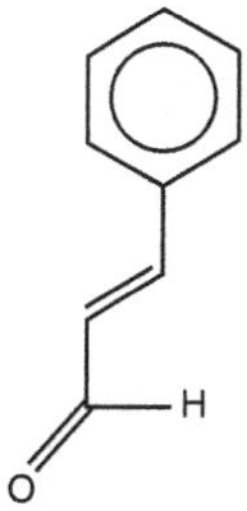

Figure 8.7 Common ketones

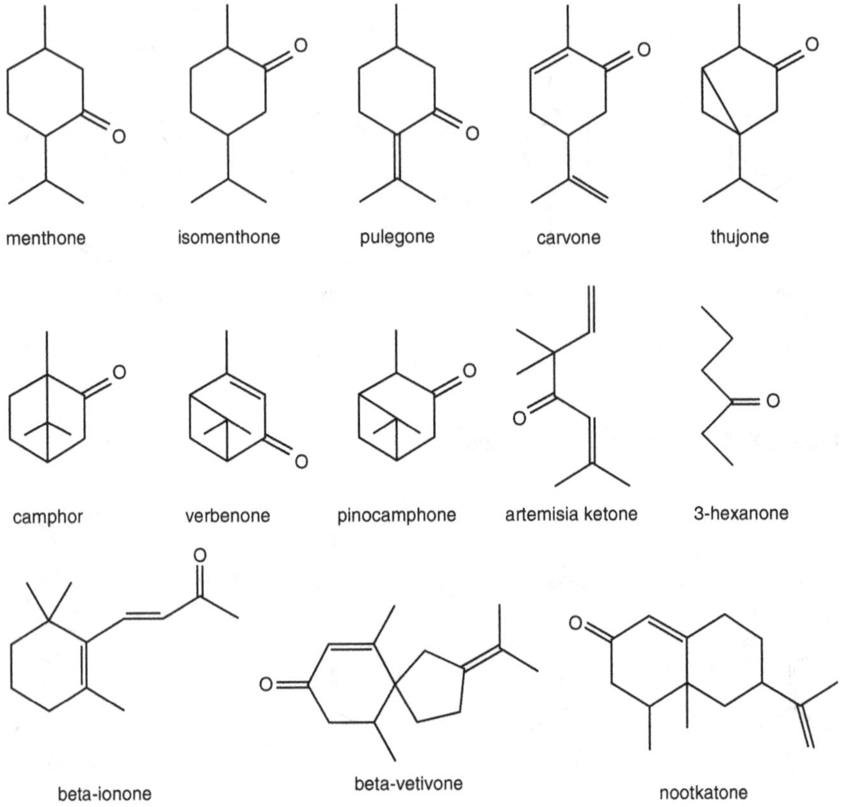

menthone   isomenthone   pulegone   carvone   thujone

camphor   verbenone   pinocamphone   artemisia ketone   3-hexanone

beta-ionone   beta-vetivone   nootkatone

Figure 8.8 Common esters

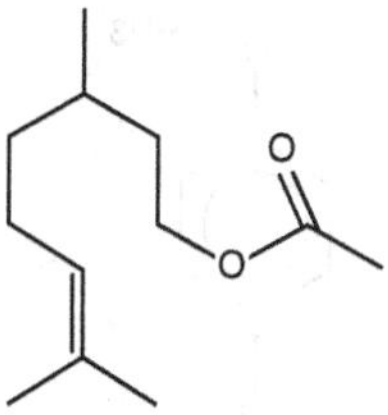

citronellyl acetate

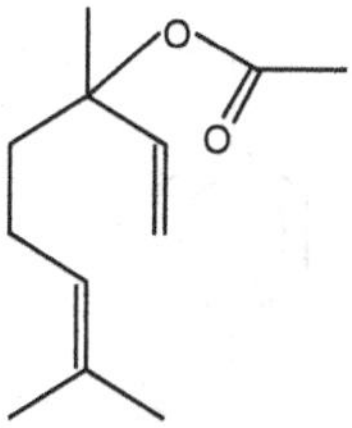

linalyl acetate

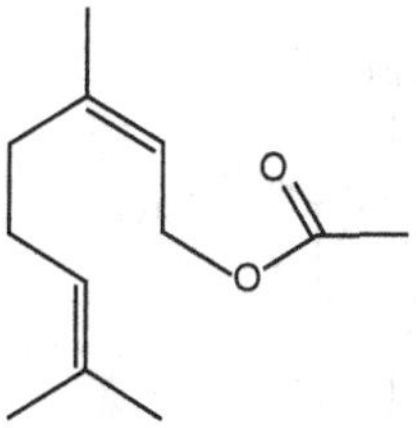

neryl acetate

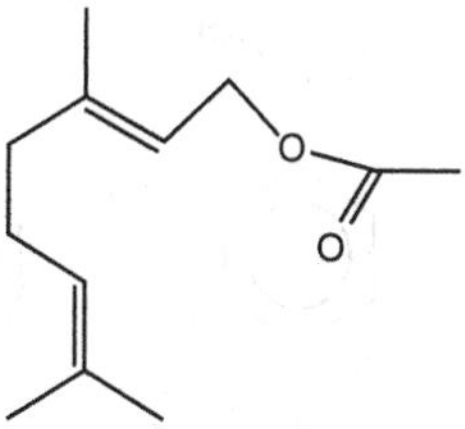

geranyl acetate

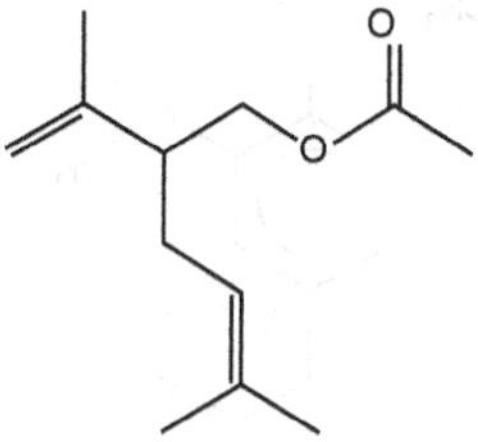

lavandulyl acetate

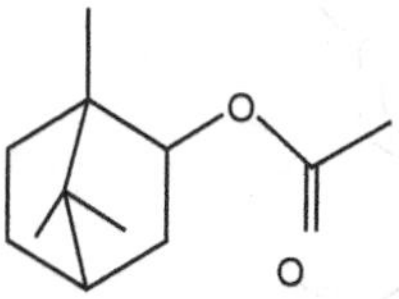

bornyl acetate

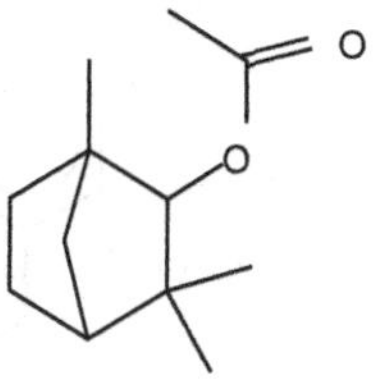

fenchyl acetate

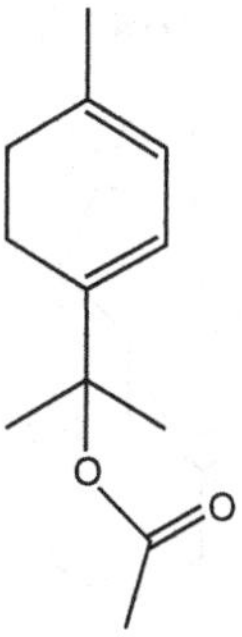

alpha-terpinyl acetate

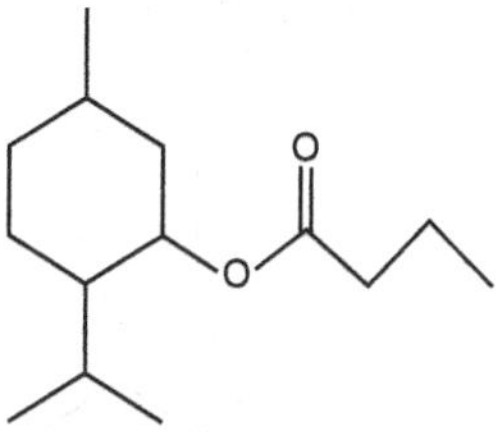

menthyl butyrate

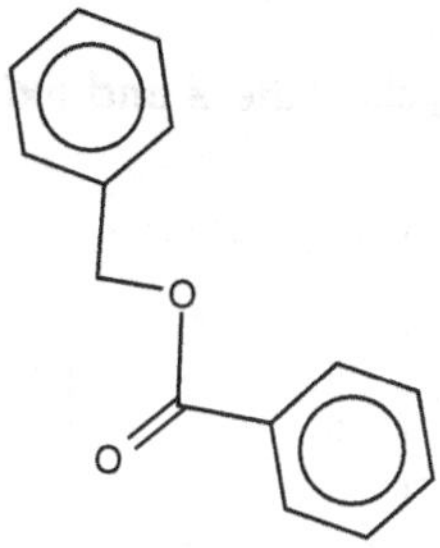

benzyl benzoate

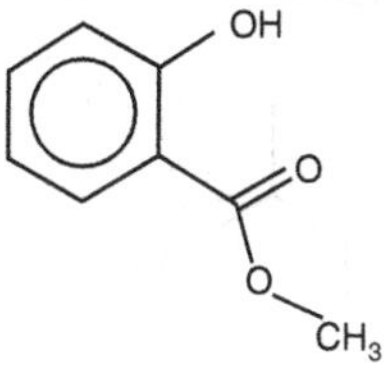

methyl salicylate

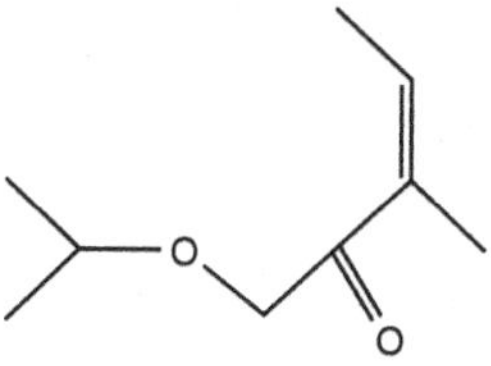

isobutyl angelate

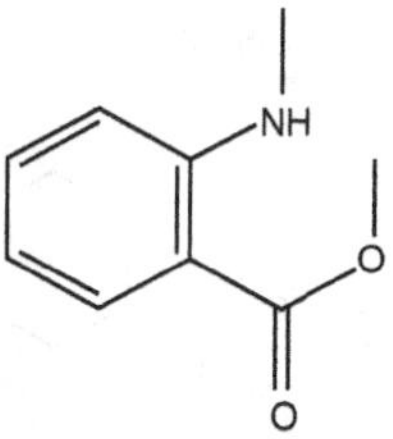

methyl N-methyl anthranilate

Figure 8.9   Common ethers

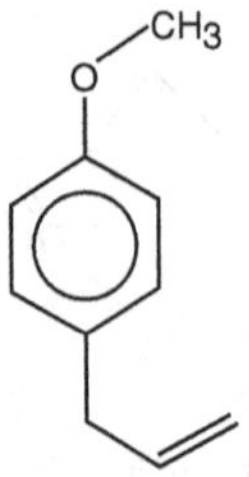

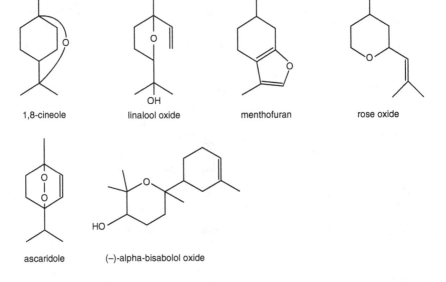

methyl chavicol 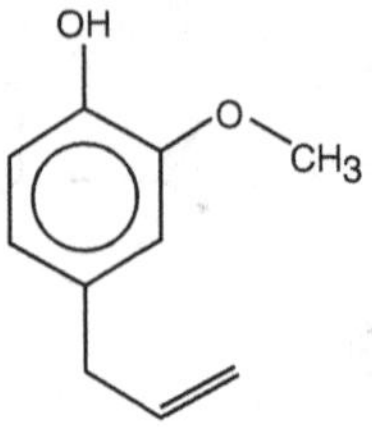   eugenol 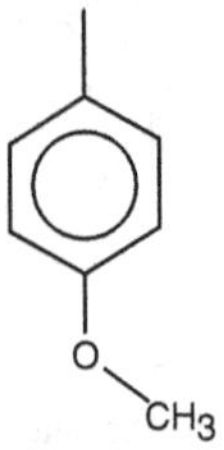   p-cresyl methyl ether    trans-anethole

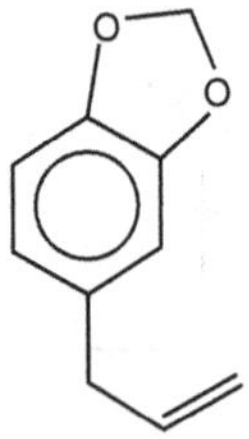

safrole 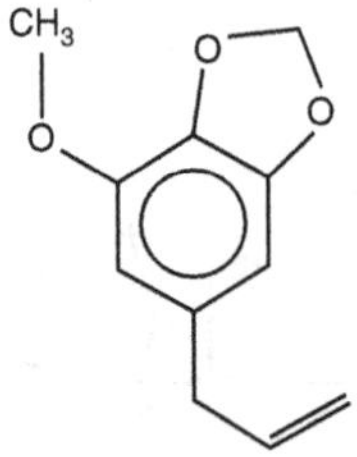   myristicin 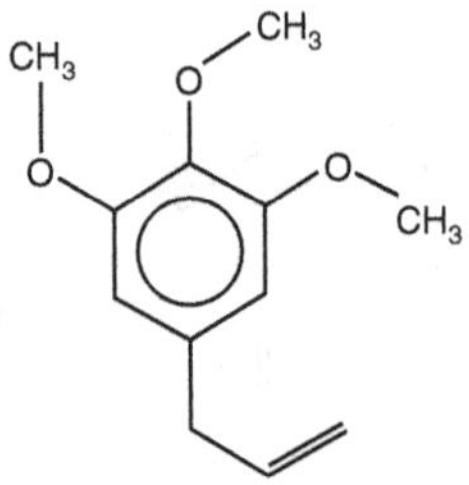   elemicin

Figure 8.10   Common cyclic ethers and oxides

1,8-cineole   linalool oxide   menthofuran   rose oxide

ascaridole   (–)-alpha-bisabolol oxide

Figure 8.11 Common lactones

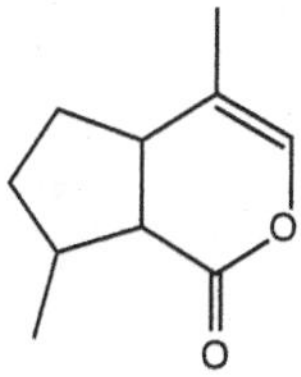

nepetalactone 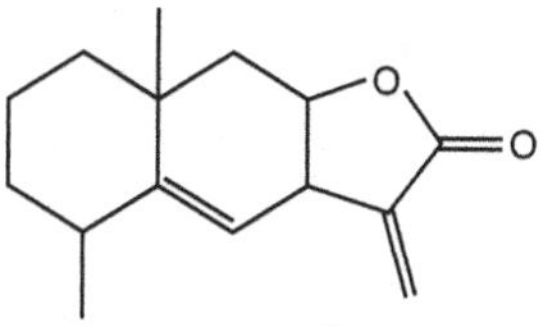 · alantolactone 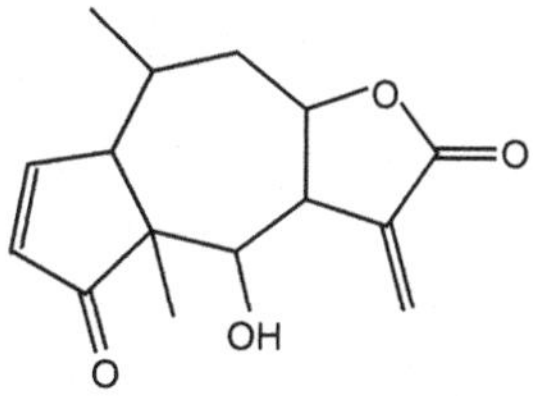 · helenalin

Figure 8.12 Common coumarins

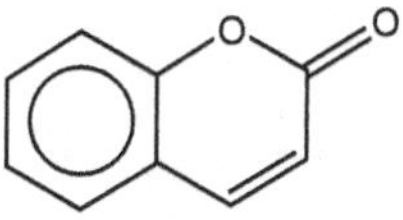

coumarin 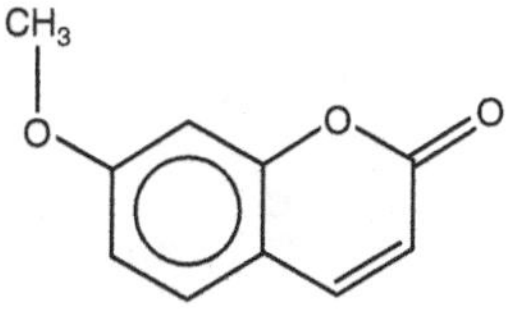 · herniarin 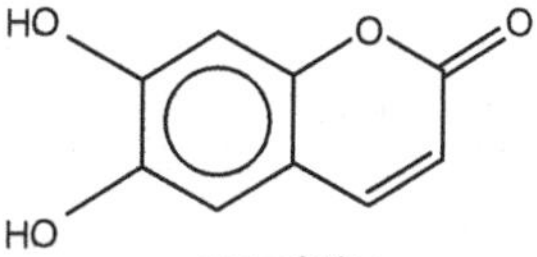 · aesculetine

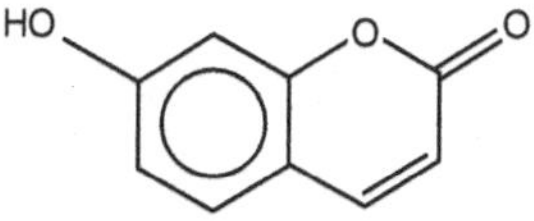

umbelliferone 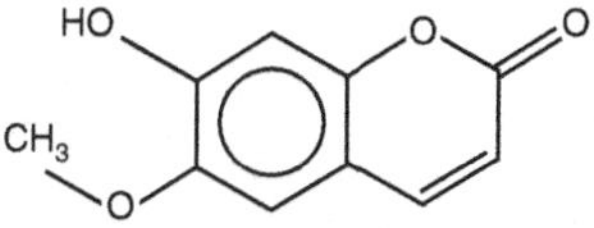 · scopoletine

Figure 8.13 Common furocoumarins

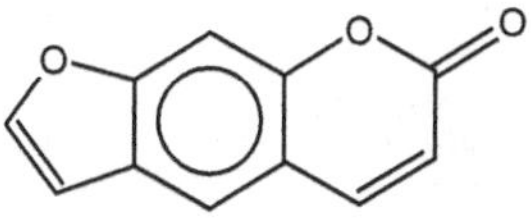

psoralen 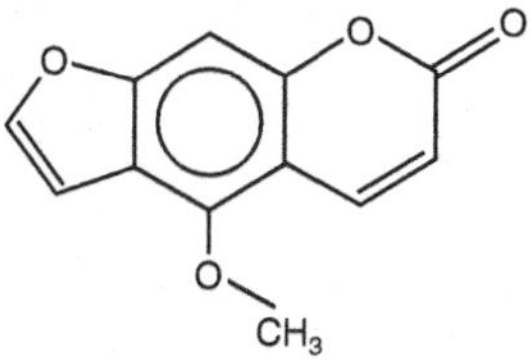 · bergapten

APPENDIX

Ready reference list of essential oils

The table following is a ready reference list of the three major constituents of each oil; these have already appeared at the end of each section in Chapters 3 and 4. 'Major' means a high percentage of the constituent is present in the oil. I have included some minor components of each oil which may, given further research, have a significant therapeutic or hazardous effect.

The information here was sourced mainly from entries in *The Complete Essential Oil Database* 1994 (Boelens Aroma Chemical Information Service, The Netherlands), and where possible I quote the original article as per that database at the end of this chapter. The percentages have been rounded to the nearest decimal point, and if I have found variations or chemotypes of the oil, I have included both (see, for example, Frankincense and Basil).

A word of warning is necessary here about the percentages. I have presented specific percentages but, in reality, each batch of essential oil will vary, and even repeated analysis of a single batch would probably have slight variations. Other aromatherapy books give ranges of percentages, but I feel this can be confusing, as it is unclear which constituents fluctuate with higher or lower levels of other constituents. At least this way you have a snapshot of one sample of an oil, which I have attempted to ensure is as representative as possible.

The regular column 'Progress in essential oils' by B.J. Lawrence in *Perfumer & Flavorist* magazine (Allured Publishing Corp., Carol Stream, IL USA), also contains interesting information on the constituents and composition of essential oils.

Major and minor constituents of essential oils

Name of oil	1	2	3	Distinctive minor constituents
Ajowan, *Trachyspermum copticum* L. Link[1]	thymol 61%	para-cymene 15.5%	gamma-terpinene 11%	beta-pinene 3.3%, terpinen-4-ol 1.1%, carvacrol 0.6%, alpha-pinene 0.3%
Angelica (root), *Angelica archangelica*[2]	alpha-pinene 25%	1,8-cineole 14.5%	alpha-phellandrene 13.5%	borneol 1%, ambrettolide 0.3%, pentadecanolide 0.2%, other musk-like compounds
Anise (Spain), *Pimpinella anisum*[3]	(E)-anethole 96%	limonene 0.6%	anisaldehyde 0.6%	coumarins 1% (Pénoël), anisalcohol 0.4%, (Z)-anethole 0.4%, methyl chavicol 0.3%, linalool 0.3%
Anise, Star (China), *Illicium verum* Hook. F.[4]	(E)-anethole 71.5%	foeniculin 14.5% (an azulene)	methyl chavicol 5%	limonene 1.4%, linalool 0.6%, nerolidol 0.5%, cinnamyl acetate 0.2%, (Z)-anethole trace
Annual Wormwood (Yugoslavia), *Artemisia alba*[5]	artemisia ketone 44.8%	1,8-cineole 9.6%	camphor 6.3%	alpha-ylangene 6.2%, verbenone 2.6%, artemisia alcohol 1.8%, methyl chamazulene 0.3%, benzyl isovalerate 0.3%, ethyl 2-methyl butyrate 0.2%, arteannuin B 0.2%
Artemisia alba (Belgium)[6]	isopinocamphone 34.6%	camphor 21.1%	1,8-cineole 5.7%	myrtenol 3.8%, T-cadinol 3.4%, myrtenal 3%, pinocamphone 1.5%
Basil (Comoro Islands), *Ocimum basilicum*[7]	methyl chavicol 85%	1,8-cineole 3.25%	para-cymene 2.7%	methyl eugenol 1.3%, linalool 0.96%, eugenol 0.5%, p-methoxycinnamaldehyde 0.4%, spathulenol 0.1%
Basil (Portugal), *Ocimum basilicum*[8]	linalool 38.2%	methyl chavicol 16.4%	beta-caryophyllene 7%	eugenol 5.1%, methyl (E)-cinnamate 4.7%, alpha-terpinyl acetate 4.5%, 1,8-cineole 3.5%, terpinen-4-ol 2.5%, gamma-cadinene 2.3%
Bay (West Indian), *Pimenta racemosa* Mill. JS Moore[9]	eugenol 56%	chavicol 21.6%	myrcene 13%	linalool 1.7%, methyl cinnamate 1.1%, 1-octen-3-ol 1%, 3-octanol 0.6%, terpinen-4-ol 0.3%, 1,8-cineole 0.2%

Bergamot (Calabria), *Citrus bergamia*[10]	limonene 38.4%	linalyl acetate 28%	linalool 8%	gamma-terpinene 8%, beta-pinene 7%, alpha-pinene 1.4%, bergamotene 0.3%, geranyl acetate 0.4%
Black Pepper (Sarawak), *Piper nigrum*[11]	beta-caryophyllene 34.6%	delta-3-carene 16%	limonene 14.5%	beta-pinene 7.7%, alpha-pinene 3.7%, alpha-copaene 2.9%, delta-elemene 1.8%, cadinol 0.9%, alpha-cubebene 0.2%, torreyol 0.2%
Buchu, *Barosma betulina* Bertl. (also *Agathosma betulina*)[12]	(–)-menthone 35%	diosphenol 12%	(-)-pulegone 11%	limonene 10%, piperitone oxide 9.5%, (+)-menthone 9%, (+)-isopulegone 3%, 8-mercapto-p-3-menthanone 1%
Cajuput, *Melaleuca leucadendron* L.[13]	1,8-cineole 41.1%	alpha-terpineol 8.7%	para-cymene 6.8%	terpinolene 4.6%, gamma-terpinene 4.6%, limonene 4.1%, linalool 3.6%, eudesmols 1.9%, terpinen-4-ol 1.5%, guaiol 1.2%, beta-pinene 0.8%
Camphor (Japan), *Cinnamomum camphora* L. Nees et Eberm.[14] (probably Yellow or Brown camphor)	camphor 51.5%	safrole 13.4%	1,8-cineole 4.8%	piperitone 2.4%, beta-caryophyllene 1.4%, cinnamyl alcohol 0.2%, furfural 0.2%, eugenol 0.1%, cinnamaldehyde 0.1%
Caraway (Netherlands), *Carum carvi* L.[15]	(+)-carvone 50%	limonene 46%	cis-dihydrocarvone 0.5%	myrcene 0.4%, carveol 0.4%, perillaldehyde 0.1%
Cardamom (Reunion), *Elettaria cardamomum* L. Maton var. *alpha-minor*[16]	1,8-cineole 48.4%	alpha-terpinyl acetate 24%	limonene 6%	alpha-terpineol 1.9%, geraniol 1.6%, terpinen-4-ol 1.4%, linalool 1.2%, borneol 0.3%, geranial 0.3%, linalyl acetate 0.2%, carvone 0.2%, neral 0.2%
Catmint, *Nepeta cataria*[17]	nepetalactone 1 80.5%	nepetalactone 2 10%	beta-caryophyllene 4.4%	alpha-humulene 0.8%, caryophyllene oxide 0.6%, beta-farnesene 0.2%

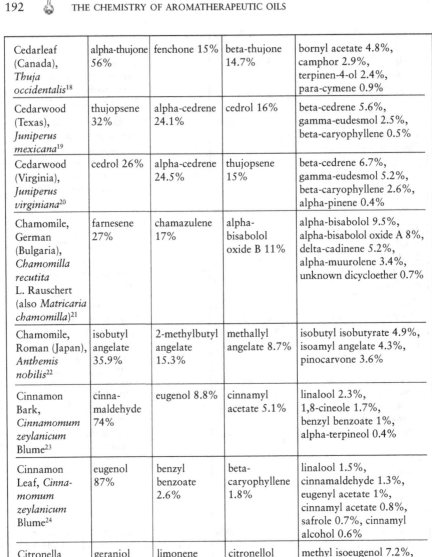

Cedarleaf (Canada), *Thuja occidentalis*[18]	alpha-thujone 56%	fenchone 15%	beta-thujone 14.7%	bornyl acetate 4.8%, camphor 2.9%, terpinen-4-ol 2.4%, para-cymene 0.9%
Cedarwood (Texas), *Juniperus mexicana*[19]	thujopsene 32%	alpha-cedrene 24.1%	cedrol 16%	beta-cedrene 5.6%, gamma-eudesmol 2.5%, beta-caryophyllene 0.5%
Cedarwood (Virginia), *Juniperus virginiana*[20]	cedrol 26%	alpha-cedrene 24.5%	thujopsene 15%	beta-cedrene 6.7%, gamma-eudesmol 5.2%, beta-caryophyllene 2.6%, alpha-pinene 0.4%
Chamomile, German (Bulgaria), *Chamomilla recutita* L. Rauschert (also *Matricaria chamomilla*)[21]	farnesene 27%	chamazulene 17%	alpha-bisabolol oxide B 11%	alpha-bisabolol 9.5%, alpha-bisabolol oxide A 8%, delta-cadinene 5.2%, alpha-muurolene 3.4%, unknown dicycloether 0.7%
Chamomile, Roman (Japan), *Anthemis nobilis*[22]	isobutyl angelate 35.9%	2-methylbutyl angelate 15.3%	methallyl angelate 8.7%	isobutyl isobutyrate 4.9%, isoamyl angelate 4.3%, pinocarvone 3.6%
Cinnamon Bark, *Cinnamomum zeylanicum* Blume[23]	cinnamaldehyde 74%	eugenol 8.8%	cinnamyl acetate 5.1%	linalool 2.3%, 1,8-cineole 1.7%, benzyl benzoate 1%, alpha-terpineol 0.4%
Cinnamon Leaf, *Cinnamomum zeylanicum* Blume[24]	eugenol 87%	benzyl benzoate 2.6%	beta-caryophyllene 1.8%	linalool 1.5%, cinnamaldehyde 1.3%, eugenyl acetate 1%, cinnamyl acetate 0.8%, safrole 0.7%, cinnamyl alcohol 0.6%
Citronella (Ceylon), *Cymbopogon nardus* L. Rendle[25]	geraniol 18%	limonene 9.7%	citronellol 8.4%	methyl isoeugenol 7.2%, borneol 6.6%, citronellal 5.2%, geranyl formate 4.2%, citronellyl acetate 1.9%, methyl eugenol 1.7%, geranyl butyrate 1.5%
Citronella (Java), *Cymbopogon winterianus* Jowitt[26]	citronellal 36.8%	geraniol 21.4%	citronellol 15%	citronellyl acetate 2.5%, elemol 2.1%, isopulegol 2.1%, geranial 0.8%, neral 0.6%, eugenol 0.6%, alpha-bergamotene 0.1%

Clary sage (USA), *Salvia sclarea*[27]	linalyl acetate 49%	linalool 24%	germacrene D 3%	alpha-terpineol 3%, geranyl acetate 2.5%, geraniol 2.4%, sclareol 0.3%, 3-hexenol 0.2%, spathulenol trace
Clove Bud (Madagascar), *Eugenia caryophyllus* (Spreng.) Bullock[28]	eugenol 76.6%	beta-caryophyllene 9.8%	eugenyl acetate 7.6%	alpha-humulene 1.2%, isoeugenol 0.2%, alpha-terpinyl acetate 0.2%, alpha-copaene 0.2%, alpha-cubebene 0.2%, methyl benzoate 0.1%
Cypress (Croatia), *Cupressus sempervirens*[29]	alpha-pinene 47.9%	delta-3-carene 19.8%	cedrol 6.8%	limonene 4.1%, alpha-terpinenyl acetate 1.7%, sabinene 1.2%, bornyl acetate 0.4%, delta-cadinene 0.3%
Elemi, *Canarium luzonicum*[30]	limonene 54%	alpha-phellandrene 15.1%	elemol 15%	elemicin 3.5%, 1,8-cineole 2.5%, myrcene 2.4%, methyl eugenol 0.3%, carvone 0.2%
Eucalyptus (Spain), *Eucalyptus globulus*[31]	1,8-cineole 66.1%	alpha-pinene 14.7%	limonene 3%	aromadendrene 2.2%, pinocarvone 1%, alpha-terpinyl acetate 0.9%, globulol 0.5%, alpha-terpineol 0.4%, beta-pinene 0.4%, 3-methylbutanal 0.2%, terpinen-4-ol 0.1%, isoamyl isovalerate trace
Fennel, Sweet (Turkey), *Foeniculum vulgare* Mill. var. *dulce*[32]	(E)-anethole 80%	limonene 6%	methyl chavicol 4.5%	fenchone 2%, anisaldehyde 1%, 1,8-cineole 0.4%, (Z)-anethole 0.3%, carvone 0.1%, citral 0.1%, octanal 0.1%
Fir Balsam (Canada), *Abies balsamea*[33]	beta-pinene 30%	delta-3-carene 21.5%	bornyl acetate 11.9%	limonene 11%, borneol 0.4%, piperitone 0.2%, terpinen-4-ol 0.1%
Frankincense (Somalia), *Boswellia carterii*[34]	alpha-pinene 34.5%	alpha-phellandrene 14.6%	para-cymene 14%	1,8-cineole 1%, beta-bourbenene 0.3%, beta-elemene 0.1%, beta-caryophyllene 0.1%, allo-aromadendrene trace
Frankincense (Somalia), *Boswellia carterii*[35]	octyl acetate 60%	1-octanol 12.7%	alpha-pinene 3.5%	incensol 2.7%, linalool 2%, isocembrene 1.8%, 1,8-cineole 1.6%, bornyl acetate 1.1%, decyl acetate 0.3%

Geranium, Bourbon, *Pelargonium graveolens*[36]	citronellol 21.2%	geraniol 17.5%	linalool 12.9%	citronellyl formate 8.4%, geranyl formate 7.6%, isomenthone 7.2%, guaia-6,9-diene 3.9%, menthone 1.5%, geranyl butyrate 1.3%, citronellyl butyrate 1.3%, cis-rose oxide 0.6%, furopelargone A 0.4%, 2-phenylethyl tiglate 0.4%, trans-rose oxide 0.2%
Ginger (China), *Zingiber officinale* Roscoe[37]	ar-curcumene 16.3%	alpha-zingiberene 14.2%	beta-sesquiphellan-drene 10.6%	bisabolenes 10.7%, citral 4%, geraniol 1.7%, zingiberenol 1.4%, citronellol 1.2%, beta-eudesmol 0.9%, 2-undecanone 0.3%, 6-methyl-5-hepten-2-one 0.2%, hexanal 0.1%, decanal trace
Grapefruit (Israel), *Citrus paradisi*[38]	limonene 93%	myrcene 1.97%	alpha-pinene 0.6%	nootkatone 0.3%, octanal 0.3%, decanal 0.3%
Ho Leaf, *Cinnamomum camphora* Sieb. ssp. *formosana* var. *brientalis*[39]	linalool 95%	camphor 0.4%	limonene 0.2%	linalyl acetate 0.2%, 1,8-cineole 0.1%, alpha-terpineol 0.1%
Hyssop (France), *Hyssopus officinalis*[40]	isopino-camphone 32.6%	beta-pinene 22.9%	pinocamphone 12.2%	methyl myrtenate 4.8%, myrtenol methyl ether 2.7%, spathulenol 2.2%, elemol 1.7%, methyl chavicol 1.3%, allo-aromadendrene 0.5%, methyl eugenol 0.5%
Immortelle/ Everlasting, *Helichrysum italicum* G. Don.[41]	alpha-pinene 21.7%	gamma-curcumene 10.4%	italidiones 8%	neryl acetate 6.1%, beta-selinene 6%, beta-caryophyllene 5%, alpha-curcumene 4%, italicene 4%, alpha selinene 3.6%, bisabolane hydroxyketones 2%, isoitalicene 1.5%, neryl propionate 1.2%, 2-methylbutyl angelate 0.6%, nerolidol 0.3%, borneol 0.2%, italicene epoxide 0.2%, italicene ethers 0.2%

Jasmine (France), *Jasminum grandiflorum*[42]	benzyl acetate 22%	benzyl benzoate 14.5%	phytyl acetate 10.2%	linalool 6.4%, methyl cis-jasmonate 3%, (Z)-3-hexenyl benzoate 2.6%, cis-jasmone 2.6%, indole 2.4%, methyl anthranilate 2%, benzyl alcohol 1.6%
Jasmine (Malati), *Jasminum sambac* L. Aiton[43]	linalool 25%	(-)-germacra-1,6-dien-5-ol 20%	alpha-farnesene 12.5%	phytol 8%, (Z)-3-hexenyl benzoate 5%, benzyl acetate 3%, benzyl alcohol 3%, methyl anthranilate 3%, benzyl benzoate 0.5%, indole 0.1%, jasmone 0.1%, p-cresol 0.1%, eugenol 0.1%, methyl jasmonate trace
Juniper berry, *Juniperus communis*[44]	alpha-pinene 33%	myrcene 11%	beta-farnesene 10.5%	gamma-elemene 2.9%, beta-caryophyllene 2.7%, beta-pinene 2.5%, sabinene hydrate 0.9%, aromadendrene 0.6%, bornyl acetate 0.4%, verbenone 0.2%
Kunzea, *Kunzea ambigua* (Smith) Druce[45]	Alpha-pinene 39.9%	1,8-cineole 15.8%	globulol 11.9%	viridiflorol 9.4%, bicyclogermacrene 5.1%, alpha-terpineol 2.9%, traces (<0.05%) of calamenene, spathulenol, citronellol, ledol and limonene
Lavandin (Abrial quality, France), *Lavandula hybrida* Rev. (*L. x intermedia* Emeric ex Loiseleur)[46]	linalool 33.5%	linalyl acetate 27.1%	camphor 9.5%	1,8-cineole 8.1%, borneol 2.5%, beta-caryophyllene 2.4%, lavandulyl acetate 1.7%, lavandulol 0.9%, 1-octenyl-3-acetate 0.5%, 3-octanone 0.4%, 1-octen-3-ol 0.3%, hexyl butyrate 0.3%
Lavender (France), *Lavandula angustifolia* Mill.[47]	linalyl acetate 40%	linalool 31.5%	(Z)-beta-ocimene 6.7%	beta-caryophyllene 5.2%, lavandulyl acetate 4.2%, terpinen-4-ol 4%, 3-octanone 1.5%, lavandulol 0.7%, 1,8-cineole 0.7%, camphor 0.3%
Lavender, Spike (Spain), *Lavandula latifolia* Medicus[48]	1,8-cineole 36.3%	linalool 30.3%	camphor 8%	borneol 2.8%, alpha-terpineol 2.6%, caryophyllene oxide 2.4%, coumarin 2.4%, linalool oxide 0.5%, isoborneol 0.3%

Lemon (Argentina), *Citrus limonum*[49]	limonene 70%	beta-pinene 11%	gamma-terpinene 8%	citral 1.6%, trans-alpha-bergamotene 0.4%, geranyl acetate 0.2%, nonanal 0.1%
Lemongrass, *Cymbopogon flexuosus*[50]	geranial 47%	neral 28%	farnesol 13%	Alpha-terpineol 2.3%, borneol 1.9%, geraniol 1%
Lime, Persian, *Citrus latifolia* Tanaka[51]	limonene 58%	gamma-terpinene 16%	beta-pinene 6%	alpha-terpineol 2.2%, p-cymene 1.6%, 1,8-cineole 1%, terpinen-4-ol 0.5%, geranial 0.1%
Mandarin (Italy), *Citrus deliciosa* Tenore[52]	limonene 71%	gamma-terpinene 18.5%	alpha-pinene 2.4%	alpha-sinensal 0.2%, octanal 0.2%, methyl N-methyl anthranilate 0.2%, decanal 0.1%, nonanal 0.1%
Marjoram, Sweet, *Marjorana hortensis* Moench (*Origanum majorana*)[53]	terpinen-4-ol 36.3%	cis-sabinene hydrate 15.9%	para-cymene 9.5%	alpha-terpineol 8.2%, linalool 3.9%, linalyl acetate 3.5%, bicyclogermacrene 2.5%, beta-caryophyllene 2%
May Chang, *Litsea cubeba* (berry)[54]	geranial 40%	neral 33.8%	limonene 8.3%	6-methyl-5-hepten-2-one 4.4%, linalool 1.7%, linalyl acetate 1.6%, geraniol 1.5%, alpha-terpinyl acetate 0.2%
Melissa (Lemon balm), *Melissa officinalis* L.[55]	geranial 45%	neral 35%	6-methyl-5-hepten-2-one 3%	beta-caryophyllene 2%, citronellal 2%, geranyl acetate 2%, aesculetine 0.5% (Pénoël), damascone 0.1%, eugenol 0.1%, vetispirane 0.1%
Mugwort (Germany), *Artemisia absinthum*[56]	beta-thujone 46%	sabinyl acetate 25%	trans-sabinol 3.2%	lavandulyl acetate 2.7%, alpha-thujone 2.7%, geranyl propionate 1.4%, 1,8-cineole 0.7%, lavandulol 0.5%, terpinen-4-ol 0.4%, gamma-cadinene 0.3%
Myrrh gum (headspace), *Commiphora myrrha* (Nees) Engler[57]	delta-elemene 28.7%	alpha-copaene 10%	beta-elemene 6.1%	methyl isobutyl ketone 5.6%, 2-methyl-5-isopropenyl furan 4.6%, 3-methyl-2-butenal 2.2%, 2-methylfuran 1.9%, dihydrocurzerenone 1.1%, 4,4-dimethyl-2-butenolide 1%

Myrtle (Spain, wild), *Myrtus communis* L.[58]	myrtenyl acetate 35.9%	1,8-cineole 29.9%	alpha-pinene 8.1%	limonene 7.5%, alpha-terpineol 4.1%, methyl eugenol 2.3%, carvacrol 0.6%, myrtenol 0.6%, linalyl acetate 0.5%, isobutyl isobutyrate 0.4%
Neroli bigarade, *Citrus aurantium* var. *amara*[59]	linalool 37.5%	limonene 16.6%	beta-pinene 11.8%	geraniol 4.3%, linalyl acetate 2.8%, nerolidol 2.6%, geranyl acetate 1.7%, neryl acetate 1%, farnesol 0.9%, terpinen-4-ol 0.8%, alpha-terpinyl acetate 0.2%, indole 0.1%, methyl N-methyl anthranilate 0.1%, cis-jasmone 0.1%
Niaouli (Madagascar), *Melaleuca quinquenervia* Cav.[60]	1,8-cineole 41.8%	viridiflorol 18.1%	limonene 5.5%	beta-caryophyllene 5%, alpha-pinene 5%, alpha-terpineol 5%, ledol 2.5%, caryophyllene oxide 0.6%, (E)-nerolidol 0.4%
Nutmeg, *Myristica fragrans*[61]	alpha-pinene 22%	sabinene 18.6%	beta-pinene 15.6%	terpinen-4-ol 7.9%, myristicin 6%, gamma-terpinene 5.1%, safrole 2%, alpha-terpineol 1%, eugenol 0.2%, isoeugenol 0.2%
Orange, Sweet (Brazil), *Citrus sinensis*[62]	limonene 89%	myrcene 1.71%	beta-bisabolene 1.29%	1-nonanol 0.7%, linalool 0.4%, neral 0.3%, decanal 0.2%, geranial 0.2%, linalyl acetate 0.1%, auraptene 0.1% (Pénoël)
Oregano (Greek), *Origanum vulgare* L. ssp. *viride* (Boiss.)[63]	thymol 85.6%	carvacrol 4.3%	gamma-terpinene 2.7%	beta-caryophyllene 2.3%, para-cymene 2.3%, camphor 0.1%, 1,8-cineole 0.1%
Palmarosa (India), *Cymbopogon martinii* Stapf. var. *motia*[64]	geraniol 80%	geranyl acetate 8.3%	linalool 2.8%	beta-caryophyllene 1.8%, farnesol 1%, neral 0.4%, alpha-farnesene 0.3%, gamma-selinene 0.2%, geranyl butyrate 0.2%
Patchouli (Indonesia), *Pogostemon cablin*, Benth.[65]	patchouli alcohol 33%	alpha-patchoulene 22%	beta-caryophyllene 20%	beta-patchoulene 13%, beta-elemene 6%, norpatchoulenol 1%, pogostol 0.4%, caryophyllene oxide 0.3%, delta-guaiene 0.3%, patchoulenone 0.1%, patchouli oxide 0.1%, patchouli pyridine 0.1%

Pennyroyal, *Mentha pulegium*[66]	(+)-pulegone 63.5%	(+)-isomenthone 19.7%	(+)-neoisomenthol 5.7%	3-octanol 2%, (−)-isopulegone 0.7%, limonene 0.7%, piperitenone 0.6%, 3-octyl actate 0.1%, menthofuran trace
Peppermint (USA), *Mentha piperita* L. var. Mitcham[67]	(−)-menthol 42.8%	menthone 19.4%	sabinene hydrate 6.6%	1,8-cineole 5.2%, neomenthol 4.2%, isomenthone 3.2%, beta-caryophyllene 2.3%, menthofuran 2%, limonene 1.6%, pulegone 0.9%
Petitgrain bigarade, *Citrus aurantium* L. ssp. *amara*[68]	linalyl acetate 45.5%	linalool 24.1%	alpha-terpineol 5.2%	nerol 8%, geranyl acetate 4.2%, limonene 4%, neryl acetate 2.2%, geraniol 1.8%, beta-caryophyllene 1.6%, 2-phenylethanol 0.2%, methyl N-methyl anthranilate 0.1%, indole 0.1%, 2-methoxy-3-isobutyl pyrazine trace
Pine, *Pinus pinaster*[69]	alpha-pinene 44.1%	beta-pinene 29.5%	myrcene 4.7%	beta-caryophyllene 3.5%, delta-3-carene 3.3%, alpha-terpineol 1.4%, alpha-humulene 0.5%
Pine, dwarf, *Pinus mugo* ssp. *pumilia*[70]	delta-3-carene 35%	beta-phellandrene 15%	alpha-pinene 13.1%	beta-caryophyllene 3.3%, bornyl acetate 1.3%, p-cymene 0.7%, germacrene D 0.5%, T-cadinol trace
Pine, Scotch, *Pinus sylvestris*[71]	alpha-pinene 42%	delta-3-carene 20.5%	limonene 5.2%	cadinene 4.8%, germacra-1-(10)-E,5E-dien-4-ol 1.9%, T-cadinol 0.6%, bornyl acetate 0.1%
Rose (Bulgaria), *Rosa damascena* Mill. (otto)[72]	citronellol 33.4%	stearopten waxes 24%	geraniol 18%	nerol 5.9%, linalool 2.1%, eugenol 1.5%, 2-phenyl ethyl alcohol 1.3%, farnesol 0.9%, methyl eugenol 0.5%, geranial 0.5%, cis-rose oxide 0.4%, trans-rose oxide 0.4%, 3-hexenal 0.3%, carvone 0.2%, beta-damascenone 0.1%
Rose (Egypt), *Rosa damascena* Mill.[73]	2-phenyl ethyl alcohol 37.9%	geraniol 15.8%	citronellol 12.6%	farnesol 6.3%, nerol 4%, linalool 2.2%, eugenol 1.2%, alpha-ionone 1%

Rosemary (Spain) (camphor),[74] *Rosmarinus officinalis*	alpha-pinene 22%	camphor 17%	1,8-cineole 17%	verbenone 4%, borneol 2%, bornyl acetate 1.5%, terpinen-4-ol 1.5%, alpha-terpineol 1.5%
Rosemary (Tunisia) (cineole),[75] *Rosmarinus officinalis*	1,8-cineole 51.3%	camphor 10.6%	alpha-pinene 10%	borneol 7.7%, alpha-terpineol 3.9%, terpinen-4-ol 1%, bornyl acetate 0.8%, alpha-humulene 0.7%, verbenone 0.1%
Rosewood (Brazil), *Aniba rosaeodora*[76]	linalool 85.3%	alpha-terpineol 3.5%	cis-linalool oxide 1.5%	trans-linalool oxide 1.3%, 1,8-cineole 1%, geranyl acetate 0.1%
Rue (Egypt), *Ruta graveolens*[77]	2-undecanone 49.2%	2-nonanone 24.7%	2-nonyl acetate 6.2%	bergapten 7%, limonene 6%, 2-decanone 2.8%, pregeijerene 2.1%, 2-dodecanone 1.1%
Sage (Dalmatia), *Salvia officinalis*[78]	alpha-thujone 37.1%	beta-thujone 14.2%	camphor 12.3%	1,8-cineole 12%, alpha-pinene 3.9%, alpha-humulene 3.8%, bornyl acetate 0.9%, terpinen-4-ol 0.2%
Sandalwood (India), *Santalum album*[79]	cis-alpha-santalol 50%	cis-beta-santalol 20.9%	epi-beta-santalol 4.1%	alpha-santalal 2.9%, cis-lanceol 1.7%, santalenes 1.7%, trans-beta-santalol 1.5%, spirosantalol 1.2%, cis-nuciferol 1.1%, beta-santalal 0.6%, eka-santalals 0.1%
Savory, Summer (Italy), *Satureja hortensis* L.[80]	carvacrol 48%	gamma-terpinene 28%	para-cymene 7%	alpha-terpinene 2.8%, beta-caryophyllene 1.3%, alpha-pinene 1.2%, beta-bisabolene 0.7%, methyl chavicol 0.2%
Spearmint (Greece), *Mentha spicata*[81]	(–)-carvone 42.8%	dihydrocarvone 15.7%	1,8-cineole 5.8%	perillyl alcohol 4.5%, alpha-terpinenyl acetate 4.5%, beta-bourbenene 2.6%, beta-caryophyllene 2.5%
Tansy (Belgium), *Tanacetum vulgare*[82]	beta-thujone 50%	trans-chrysanthemyl acetate 20%	camphor 6.4%	germacrene D 5%, alpha-thujone 1.8%, 1,8-cineole 0.5%, alpha-pinene 0.5%, eugenol 0.4%
Tarragon (USA), *Artemisia dranunculus*[83]	methyl chavicol 80%	beta-ocimenes 14%	limonene 2.5%	alpha-pinene 0.5%, methyl eugenol 0.5%, eugenol 0.2%, elemicin 0.1%

Tea-tree, *Melaleuca alternifolia*[84]	terpinen-4-ol 45.4%	gamma-terpinene 15.7%	alpha-terpinene 7.1%	para-cymene 6.2%, alpha-terpineol 5.3%, 1,8-cineole 3%, alpha-pinene 2.1%, limonene 1.4%
Thyme (Italy), *Thymus vulgaris*[85]	thymol 27.4%	para-cymene 21.9%	gamma-terpinene 12%	beta-caryophyllene 3.3%, linalool 2.6%, 1,8-cineole 2.2%, carvacrol 1.1%, alpha-thujone 0.3%, delta-cadinene 0.2%
Turmeric (Indonesia), *Curcuma longa*[86]	turmerone 29.5%	ar-turmerone 24.7%	turmerol 20%	beta-curcumene 2.5%, alpha-atlantone 2.4%, curcuphenol 0.6%, beta-bisabolol 0.3%
Vanilla, *Vanilla fragrans* Ames[87]	vanillin 85% (ether)	4-hydroxy-benzaldehyde 8.5%	4-hydroxy-benzyl methyl ether 1%	not specified, but includes alkyl benzenes and esters
Vetiver, *Vetivera zizanoides* Stapf.[88]	vetiverol 50%	vetivenes 20%	alpha-vetivol 10%	vetivones 10%, vetispirenes 2%, khusimol 1%, khusimone 1%, vetiselinene 1%, vetiazulene 0.1%
Wintergreen (China), *Gaultheria procumbens*[89]	methyl salicylate 90%	safrole 5%	linalool 2%	1,8-cineole 1%, camphor 0.5%, alpha-pinene 0.5%, eugenol 0.2%, ethyl salicylate 0.1%
Yarrow, *Achille millefolium*[90]	camphor 17.7%	sabinene 12.3%	1,8-cineole 9.5%	alpha-pinene 9.4%, iso-artemisia ketone 8.6%, beta-pinene 7.3%, terpinen-4-ol 4.3%, borneol 2.5%
Ylang ylang, *Cananga odorata*[91]	linalool 19%	beta-caryophyllene 10.5%	germacrene D 10.2%	p-cresyl methyl ether 8.7%, benzyl benzoate 7.3%, geranyl acetate 6.7%, benzyl acetate 4.6%, benzyl salicylate 2%, farnesol 1.8%, cadinols 1.8%, eugenol 0.3%, 3-methyl-2-butenyl acetate 0.1%

NOTES

Introduction
1 National Library of Medicine PubMed entry to Med line database. http://www.nebi-nlm.nih.gov/PubMed/

Chapter 1
1 F. Brescia, J. Arents, H. Meislich and A. Turk (1980), *Fundamentals of Chemistry*, 4th edn, Academic Press Inc., Orlando, Florida.

Chapter 2
1 R.T. Morrison and R.N. Boyd (1987), *Organic Chemistry*, 5th edn, Allyn & Bacon Inc., Boston, p. 409.
2 J.E. Simon, A.F. Chadwick and L.E. Craker (1984), *Herbs: An Indexed Bibliography, 1971–1980: The Scientific Literature on Selected Herbs, and Aromatic and Medicinal Plants of the Temperate Zone*, Archon Books, Hamden, Connecticut.
3 D. Basker and E. Putievsky (1978), 'Seasonal variation in the yields of leaf and essential oil in some Labiatae species' *Journal of Horticultural Science*, 53(3), pp. 179–83.
4 R. Tisserand and T. Balacs (1995), *Essential Oil Safety: A Guide for Health Professionals*, Churchill Livingstone, Edinburgh, p. 186.
5 S.H. Ahmad, A.A. Malek, H.C. Gan, T.L. Abdullah and A.A. Rahman (1998), 'The effect of harvest time on the quantity and chemical composition of jasmine (*Jasminum multiflorum* L.) essential oil' *Acta Hort*, (ISHS) 454, pp. 355–64; http://www.actahort.org/books/454/454_42.htm.
6 *R & D Plan for Essential Oils and Plant Extracts 2002–2006*, Rural Industries Research and Development Corporation http://www.ridc.gov.au/pub/essentoi.html [accessed 25 May 2003].

7 B.M. Lawrence (1993), 'A planning scheme to evaluate new aromatic plants for the flavor and fragrance industries', in J. Janick and J.E. Simon (eds), *New Crops*, Wiley, New York, pp. 620–7 http://www.hort.purdue.edu/newcrop/proceedings1993/v2-620.html [accessed 15 January 2003].

Chapter 3

1 G.H. Dodd (1988), 'The molecular dimension in perfumery', in Steve Van Toller and George H. Dodd (eds), *Perfumery: The Psychology and Biology of Fragrance*, Chapman Hall, London, p. 33.

2 A.L. Lehninger (1982), *Principles of Biochemistry*, Worth Publishers, New York, p. 316.

3 R. Tisserand and T. Balacs (1995), *Essential Oil Safety: A Guide for Health Professionals*, Churchill Livingstone, Edinburgh, p. 199.

4 B.M. Hausen, J. Reichling and M. Harkenthal (1999), 'Degradation products of monoterpenes are the sensitizing agents in tea tree oil' *American Journal of Contact Dermatitis*, 10(2), pp. 68–77.

5 R. Tisserand and T. Balacs, *Essential Oil Safety*, p. 189.

6 A.F. Filipsson (1996), 'Short term inhalation exposure to turpentine: toxicokinetics and acute effects in men' *Occupational and Environmental Medicine*, 53(2), pp. 100–5.

7 D. Pénoël and P. Franchomme (1990), *L'Aromathérapie exactement*, Roger Jollois, Limoges, p. 222.

8 R. Tisserand and T. Balacs, *Essential Oil Safety*, p. 197.

9 H. Schilcher and F. Leuschner (1997), 'The potential nephrotoxic effects of essential juniper oil' *Arzneimittelforschung*, July 47(7), pp. 855–8.

10 D. Pénoël and P. Franchomme, *L'Aromathérapie*, p. 219.

11 H. Igimi, R. Tamura, F. Yamamoto, K. Toraishi, A. Kataoka, Y. Ikejiri, T. Hisatsugu and H. Shimura (1991), 'Medical dissolution of gallstones: clinical experience of d-limonene as a simple, safe, and effective solvent' *Digestive Diseases And Sciences*, 36(2), pp. 200–8.

12 H. Igimi, D. Watanabe, F. Yamamoto, S. Asakawa, K. Toraishi and H. Shimura (1992), 'A useful cholesterol solvent for the medical dissolution of gallstones' *Gastroenterologica Japonica*, 27(4), pp. 536–45.

13 M.N. Gould (1997), 'Cancer chemoprevention and therapy by monoterpenes' *Environmental Health Perspectives*, June, v.105, Suppl. 4, pp. 977–9.

14 D. Pénoël and P. Franchomme (1995), *L'Aromathérapie*, p. 219.
15 ibid., p. 220.
16 Y. Tambe, H. Tsujiuchi, G. Honda, Y. Ikeshiro and S. Tanaka (1996), 'Gastric cytoprotection of the non-steroidal anti-inflammatory sesquiterpene, beta-caryophyllene' *Planta Medica*, 62(5), pp. 469–70.
17 H. Safayhi, J. Sabieraj, E.R. Sailer and H.P. Ammon (1994), 'Chamazulene: an antioxidant-type inhibitor of leukotriene B4 formation' *Planta Medica*, 60(5), pp. 410–3.
18 A. Lehninger (1982), *Principles of Biochemistry*, Worth Publishers, New York, p. 608.
19 W. Ma, Z. Miao and M.V. Novotny (1999), 'Induction of estrus in grouped female mice (*Mus domesticus*) by synthetic analogues of preputial gland constituents' *Chemical Senses*, 24(3), pp. 289–93.

Chapter 4

1 E. Guenther (1972), *The Essential Oils*, 6 volumes, Krieger, Malabar, Flanders.
2 R. Tisserand and T. Balacs (1995), *Essential Oil Safety: A Guide for Health Professionals*, Churchill Livingstone, London, p. 182.
3 D. Pénoël and P. Franchomme (1990), *L'Aromathérapie exactement*, Roger Jollois, Limoges, p. 182.
4 S. Pattnaik, V.R. Subramanayam, M. Bapaji and C.R. Kole (1997), 'Antibacterial and antifungal activity of aromatic constituents of essential oils' *Microbios*, 89(358), pp. 39–46.
5 C.F. Carson and T.V. Riley (1995), 'Antimicrobial activity of the major components of the essential oil of *Melaleuca alternifolia*' *Journal of Applied Bacteriology*, 78(3), pp. 264–9.
6 S.S. Budhiraja, M.E. Cullum, S.S. Sioutis, L. Evangelista and S.T. Habanova (1999), 'Biological activity of *Melaleuca alternifola* (tea tree) oil component, terpinen-4-ol, in human myelocytic cell line HL-60' *Journal of Manipulative Physiological Therapeutics*, 22(7), pp. 47–53.
7 D. Pénoël and P. Franchomme, *L'Aromathérapie*, p. 158.
8 N. Galeotti, L. Di Cesare Mannelli, G. Mazzanti, A. Bartolini and C. Ghelardini (2002), 'Menthol: a natural analgesic compound' *Neuroscience Letters*, April 12, 322(3), pp. 145–8.
9 G. Buchbauer, L. Jirovetz, W. Jager, H. Dietrich and C. Plank (1991), 'Aromatherapy: evidence for sedative effects of the essential oil of lavender after inhalation' *Zeitschrift fur Natürforschung [[C]]*, Nov–Dec, 46(11–12), pp. 1067–72.
10 E. Elizabetsky, L.F. Brum and D.O. Souza (1999), 'Anticonvulsant

properties of linalool in glutamate-related seizure models' *Phytomedicine*, 6(2), pp. 107–13.

11 M. Lis-Balchin and S. Hart (1999), 'Studies on the mode of action of the essential oil of Lavender (*Lavandula angustifolia* P. Miller)' *Phytotherapy Research*, 13(6), pp. 540–2.

12 R. Carle and K. Gomaa (1992), 'The medicinal use of *Matricaria flos*' *British Journal of Phytotherapy*, 2(4), pp. 147–53.

13 U.C. Luft, R. Bychkov, M. Gollasch, V. Gross, J.B. Roullet, D.A. McCarron, C. Ried, F. Hofmann, Y. Yagil, C. Yagil, H. Haller and F.C. Luft (1999), 'Farnesol blocks the L-type Ca^{2+} channel by targeting the alpha 1C subunit' *Arteriosclerosis, Thrombosis and Vascular Biology*, Apr, 19(4), pp. 959–66.

14 P.M. Zygmunt, B. Larsson, O. Sterner, E. Vinge and E.D. Hogestatt (1993), 'Calcium antagonistic properties of the sesquiterpene T-cadinol and related substances: structure-activity studies' *Pharmacology & Toxicology*, 73(1), pp. 3–9.

15 B. Shieh, Y. Lizuka and Y. Matsubara (1981), 'Monoterpenoid and sesquiterpenoid constituents of the essential oil of Hinoki Wood (*Chamaecyparis obtusa* Sieb et Zucc. Endl.)' *J. Agric. Biol. Chem.* (Japan), v.45, pp. 1497–9.

16 K. Asakura, Y. Matsuo, T. Oshima, T. Kihara, K. Minagawa, Y. Araki, K. Kagawa, T. Kanemasa and M. Ninomiya (2000), 'Omega-Agatoxin IVA-sensitive Ca(2+) channel blocker, alpha-eudesmol, protects against brain injury after focal ischemia in rats' *European Journal of Pharmacology*, 394(1), pp. 57–65.

17 L.C. Chiou, J.Y. Ling and C.C. Chang (1997), 'Chinese herb constituent beta-eudesmol alleviated the electroshock seizures in mice and electrographic seizures in rat hippocampal slices' *Neuroscience Letters*, August 15, 231(3), pp. 171–4.

18 A. Rioja, A.R. Pizzey, C.M. Marson and N.S. Thomas (2000), 'Preferential induction of apoptosis of leukaemic cells by farnesol' *FEBS Letters*, 467(2–3), pp. 291–5.

19 L.W. Wattenberg (1991), 'Inhibition of azoxymethane-induced neoplasia of the large bowel by 3–hydroxy-3,7,11–trimethyl-1,6,10–dodecatriene (nerolidol)' *Carcinogenesis*, Jan, 12(1), pp. 151–2.

20 E.V. Lassak and I.A. Southwell (1977), 'Essential oil isolates from the Australian flora' *IFFA*, May/June, pp. 126–32.

21 F. Benencia and M.C. Courreges (1999), 'Antiviral activity of sandalwood oil against herpes simplex viruses-1 and -2' *Phytomedicine*, 6(2), pp. 119–23.

22 N.P. Lopes, M.J. Kato, E.H. Andrade, J.G. Maia, M. Yoshida,

A.R. Planchart and A.M. Katzin (1999), 'Antimalarial use of volatile oil from leaves of *Virola surinamensis* (Rol.) Warb. by Waiapi Amazon Indians' *Journal of Ethnopharmacology*, November 30, 67(3), pp. 313–19.

23 D. Pénoël and P. Franchomme, *L'Aromathérapie*, p. 154.

24 ibid., p. 154.

25 ibid., p. 154.

26 S. Consentino, C.I. Tuberoso, B. Pisano, M. Satta, V. Mascia, E. Arzedi and F. Palmas (1999), 'In-vitro antimicrobial activity and chemical composition of Sardinian *Thymus* essential oils' *Letters in Applied Microbiology*, 29(2), pp. 130–5.

27 A. Ultee, E.P. Kets and E.J. Smid (1999), 'Mechanisms of action of carvacrol on the food-borne pathogen *Bacillus cereus*' *Applied Environmental Microbiology*, 65(10), pp. 4606–10.

28 G.L. Case, L. He, H. Mo and C.E. Elson (1995), 'Induction of geranyl pyrophosphate pyrophosphatase activity by cholesterol-suppressive isoprenoids' *Lipids*, 30(4), pp. 357–9.

29 R. Tisserand and T. Balacs, *Essential Oil Safety*, p. 187.

30 D. Engelstein, J. Shmueli, S. Bruhis, C. Servadio and A. Abramovici (1996), 'Citral and testosterone interactions in inducing benign and atypical prostatic hyperplasia in rats' *Comparative Biochemistry & Physiology C—Pharmacology Toxicology and Endocrinology*, 115(2), pp. 169–77; A.A. Geldof, C. Engel and B.R. Rao (1992), 'Estrogenic action of commonly used fragrant agent citral induces prostatic hyperplasia' *Urology Research*, 20(2), pp. 139–44.

31 R. Tisserand and T. Balacs, *Essential Oil Safety*, p. 187.

32 D. Pénoël and P. Franchomme, *L'Aromathérapie*, p. 211.

33 C. Viollon and J.P. Chaumont (1994), 'Antifungal properties of essential oils and their main components upon *Cryptococcus neoformans*' *Mycopathologia*, 128(3), pp. 151–3.

34 M.L. Iersel, J.P. Ploemen, I. Struik, C. van Amersfoort, A.E. Keyzer, J.G. Schefferlie and P.J. van Bladeren (1996), 'Inhibition of glutathione S-transferase activity in human melanoma cells by alpha,beta-unsaturated carbonyl derivatives. Effects of acrolein, cinnamaldehyde, citral, crotonaldehyde, curcumin, ethacrynic acid, and trans-2–hexenal' *Chemical and Biolochemical Interactions*, 102(2), pp. 117–32.

35 R. Tisserand and T. Balacs, *Essential Oil Safety*, p. 69.

36 ibid., p. 200.

37 D. Pénoël and P. Franchomme, *L'Aromathérapie*, p. 194.

38 ibid., p. 195.

39 ibid., p. 194.

40 J.M. Xie, S.S. Greenberg and G. Longenecker (1992), 'Effects of 2–camphanone on canine portal vein blood flow and rat smooth muscle' *Gastroenterology*, 102(2), pp. 394–402.

41 D. Pénoël and P. Franchomme, *L'Aromathérapie*, p. 196.

42 ibid., p. 188.

43 R. Tisserand and T. Balacs, *Essential Oil Safety*, p. 54.

44 N. Pages, G. Fournier, G. Chamorro, M. Salazar, M. Paris and C. Boudene (1989), 'Teratological evaluation of *Juniperus sabina* essential oil in mice' *Planta Medica*, 55(2), pp. 144–6.

45 D. Pénoël and P. Franchomme, *L'Aromathérapie*, p. 183.

46 G. Buchbauer, L. Jirovetz, W. Jager, H. Dietrich and C. Plank (1991), 'Aromatherapy: evidence for sedative effects of the essential oil of lavender after inhalation' *Zeitschrift fur Natürforschung [[C]]*, Nov–Dec, 46(11–12), pp. 1067–72.

47 D. Pénoël and P. Franchomme, *L'Aromathérapie*, p. 183.

48 R. Tisserand and T. Balacs, *Essential Oil Safety*, pp. 69–72, 194.

49 ibid., p. 198.

50 D. Pénoël and P. Franchomme, *L'Aromathérapie*, p. 178.

51 S. Mills and K. Bone (2000), *Principles and Practice of Phytotherapy*, Churchill Livingstone, Edinburgh, p. 380.

52 A.A. Albuquerque, A.L. Sorenson and J.H. Leal-Cardoso (1995), 'Effects of essential oil of *Croton zehntneri*, and of anethole and estragole on skeletal muscles' *Journal of Ethnopharmacology*, 49(1), pp. 41–9.

53 C. Ghelardini, N. Galeotti and G. Mazzanti (2001), 'Local anaesthetic activity of monoterpenes and phenylpropanes of essential oils' *Planta Medica*, 67(6), pp. 564–6.

54 D. Pénoël and P. Franchomme, *L'Aromathérapie*, p. 178.

55 S. Seltzer (1992), 'Biologic properties of eugenol and zinc oxide-eugenol' *Oral Surgery Oral Medicine Oral Pathology Oral Radiology and Endodontics*, 73, pp. 729–37.

56 L.M. Day, J. Ozanne-Smith, B.J. Parsons, M. Dobbin and J. Tibballs (1997), 'Eucalyptus oil poisoning among young children: mechanisms of access and the potential for prevention' *Australian and New Zealand Journal of Public Health*, 21(3), pp. 297–302; P.R. Burkard, K. Burkhardt, D.A. Haenggeli and T. Landis (1999), 'Plant-induced seizures: reappearance of an old problem' *Journal of Neurology*, 246(8), pp. 667–70.

57 T. Darben, B. Cominos and C.T. Lee (1998), 'Topical eucalyptus oil poisoning' *Australasian Journal of Dermatology*, 39(4), pp. 265–7.
58 D. Pénoël and P. Franchomme, *L'Aromathérapie*, pp. 190–1.
59 R. Tisserand and T. Balacs, *Essential Oil Safety*, p. 193.
60 D. Pénoël and P. Franchomme, *L'Aromathérapie*, p. 189.
61 W.T. Ulmer and D. Schott (1991), 'Chronic obstructive bronchitis. Effect of Gelomyrtol forte in a placebo-controlled double-blind study' *Fortschritte der Medizin*, September 20, 109(27), pp. 547–50.
62 U.R. Juergens, M. Stober, L. Schmidt-Schilling, T. Kleuver and H. Vetter (1998), 'Anti-inflammatory effects of eucalyptol (1,8–cineole) in bronchial asthma: inhibition of arachidonic acid metabolism in human blood monocytes ex vivo' *European Journal of Medical Research*, 3(9), pp. 407–12.
63 L.A. Gordon (1999), 'Compositae dermatitis' *Australasian Journal of Dermatology*, 40(3), pp. 123–8.
64 B.C. McGeorge and M.C. Steele (1991), 'Allergic contact dermatitis of the nipple from Roman chamomile ointment' *Contact Dermatitis*, 24(2), pp. 139–40; W.G. Van Ketel (1982), 'Allergy to *Matricaria chamomilla*' *Contact Dermatitis*, 8(2), p. 143.
65 J. Joydnis-Liebert, M. Murias and E. Bloszyk (2000), 'Effect of sesquiterpene lactones on antioxidant enzymes and some drug-metabolizing enzymes in rat liver and kidney' *Planta Medica*, 66(3), pp. 199–205.
66 R. Tisserand and T. Balacs (1995), *Essential Oil Safety*, pp. 38, 42–3.
67 D. Pénoël and P. Franchomme, *L'Aromathérapie*, p. 202.
68 ibid., p. 326.
69 R.L. Mazor, I.Y. Menendez, M.A. Ryan, M.A. Fiedler and H.R. Wong (2000), 'Sesquiterpene lactones are potent inhibitors of interleukin 8 gene expression in cultured human respiratory epithelium' *Cytokine*, 12(3), pp. 239–45.
70 S. Mills and K. Bone (2000), *Principles and Practice of Phytotherapy*, p. 270.
71 G. Lyss, T.J. Schmidt, I. Merfort and H.L. Pahl (1997), 'Helenalin, an anti-inflammatory sesquiterpene lactone from Arnica, selectively inhibits transcription factor NF-kappaB' *Biological Chemistry*, 378(9), pp. 951–61.
72 S. Aydin, R. Beis, Y. Ozturk and K.H. Baser (1998), 'Nepetalactone: a new opioid analgesic from *Nepeta caesarea* Boiss' *Journal of Pharmacy and Pharmacology*, 50(7), pp. 813–7.

73 A. Lehninger (1982), *Principles of Biochemistry*, Worth Publishers, NY, p. 276.
74 L.I. Grossweiner (1984) 'Mechanisms of photosensitization by furocoumarins' *National Cancer Institute Monographs*, December, v.66, pp. 47–54.
75 D. Pénoël and P. Franchomme, *L'Aromathérapie*, p. 205.
76 J.R. Casley-Smith (1999), 'Benzo-pyrones in the treatment of lymphoedema' *International Angiology*, 18(1), pp. 31–41.
77 A. Burgos, A. Alcaide, C. Alcoba, J.M. Azcona, J. Garrido, C. Lorente, E. Moreno, E. Murillo, J. Olsina-Pavia, J. Olsina-Kissler, E. Samaniego and M. Serra (1999), 'Comparative study of the clinical efficacy of two different coumarin dosages in the management of arm lymphedema after treatment for breast cancer' *Lymphology*, 32(1), pp. 3–10.
78 S. Mills and K. Bone (2000), *Principles and Practice of Phytotherapy*, p. 485.

Chapter 5

1 B. Bryant, K. Knights and E. Salerno (2003), *Pharmacology for Health Professionals*, Mosby: Elsevier (Australia) Pty Ltd, Marrickville, NSW, p. 2.
2 N. Galeotti, C. Ghelardini, L.D. Mannelli, G. Mazzanti, L. Baghiroli and A. Bartolini (2001), 'Local anaesthetic activity of (+)- and (–)-menthol' *Planta Medica*, 67(2), pp. 174–6; C. Ghelardini, N. Galeotti, G. Salvatore and G. Mazzanti (1999), 'Local anaesthetic activity of the essential oil of *Lavandula augustifolia*' *Planta Medica*, 65(8), pp. 700–3.
3 A.A. Falk, M.T. Hagberg, A.E. Löf, E.M. Wigaeus-Hjelm and Z.P. Wang (1990), 'Uptake, distribution and elimination of alpha-pinene in man after exposure by inhalation' *Scandinavian Journal of Work Environment and Health*, 16(5), pp. 372–8.
4 A. Falk-Filipsson, A.E. Löf, M.T. Hagberg, E.W. Hjelm and Z.P. Wang (1993), 'D-limonene exposure to humans by inhalation: uptake, distribution, elimination, and effects of the pulmonary function' *J. Toxical. Environ. Health*, 38(1), pp. 77–88.
5 K.P. Svoboda, G. Ruzickova, R. Allan and J.B. Hampson (2000), 'An investigation into drop sizes of essential oils using different dropper types' *International Journal of Aromatherapy*, 10(3/4), pp. 99–103.
6 G. Buchbauer, L. Jirovetz, W. Jäger, C. Plank and H. Dietrich (1993), 'Fragrance compounds and essential oils with sedative

effects upon inhalation' *Journal of Pharmaceutical Sciences*, 82(6), pp. 660–4.

7 H. Römmelt, H. Drexel and K. Dirnägl (1978), 'Wirkstoffaufnahme aus planzlichen Badezusätzen' *Die Heilkunst*, 91, pp. 249–54.

8 B. Bryant et al., *Pharmacology for Health Professionals*, p. 107.

9 C. Kohlert, I. van Rensen, R. Marz, G. Schindler, E.U. Graefe and M. Veit (2000), 'Bioavailability and pharmacokinetics of natural volatile terpenes in animals and humans' *Planta Medica*, 66(6), pp. 495–505.

10 D. Pénoël and P. Franchomme (1990), *L'Aromathérapie exactement*, Roger Jollois, Limoges, p. 297.

11 W. Jäger, G. Buchbauer., L. Jirovetz and M. Fritzer (1992), 'Percutaneous absorption of lavender oil from a massage oil' *Journal of the Society of Cosmetic Chemists*, 43, pp. 49–54.

12 A.C. Williams and B.W. Barry (1991), 'Terpenes and the lipid-protein-partitioning theory of skin penetration enhancement' *Pharmacological Research*, 8(1), pp. 17–24.

13 P.A. Cornwell and B.W. Barry (1994), 'Sesquiterpene components of volatile oils as skin penetration enhancers for the hydrophilic permeant 5-fluorouracil' *Journal of Pharmacy & Pharmacology*, 46(4), pp. 261–9.

14 K. Zhao and J. Singh (1998), 'Mechanisms of percutaneous absorption of tamoxifen by terpenes: eugenol, d-limonene and menthone' *Journal of Controlled Release*, 55(2–3), pp. 253–60.

15 Y. Kaplun-Frischoff and E. Touitou (1997), 'Testosterone skin permeation enhancement by menthol through formation of eutectic with drug and interaction with skin lipids' *Journal of Pharmaceutical Science*, 86(12), pp. 1394–9.

16 D.A. Godwin and B. Michniak (1999), 'Influence of drug lipophilicity on terpenes as transdermal penetration enhancers' *Drug Development and Industrial Pharmacy*, 25(8), pp. 905–15.

17 A.C. Williams and B.W. Barry (1991), 'Terpenes and the lipid-protein-partitioning theory of skin penetration enhancement' *Pharmacological Research*, 8(1), pp. 17–24.

18 P.A. Cornwell and B.W. Barry (1994), 'Sesquiterpene components of volatile oils as skin penetration enhancers for the hydrophilic permeant 5-fluorouracil' *Journal of Pharmacy & Pharmacology*, 46(4), pp. 261–9.

19 J. Grassmann, D. Schneider, D. Weiser and E.E. Elstner (2001), 'Antioxidative effects of lemon oil and its components on copper induced oxidation of low density lipoprotein' *Arzneimittel-Forschung*, 51(10), pp. 799–805.

20 H.P. Rang and M.M. Dale (1987), *Pharmacology*, Churchill Livingstone, Edinburgh, p. 71.

21 A.A. Falk et al., 'Uptake, distribution and elimination of, alpha-pinene', p. 374.

22 M.D. Rawlins (1989), 'Clinical pharmacology of the skin', in Paul Turner (ed.), *Recent Advances in Clinical and Pharmacology and Toxicology*, Churchill Livingstone, Edinburgh, pp. 121–35.

23 H.P. Rang and M.M. Dale, *Pharmacology*, p. 76.

24 W. Jager, M. Mayer, G. Reznicek and G. Buchbauer (2001), 'Percutaneous absorption of the monoterperne carvone: implication of stereoselective metabolism on blood levels' *Journal of Pharmacy & Pharmacology*, 53(5), pp. 637–42.

25 I.R. Hardcastle, M.G. Rowlands, A.M. Barber, R.M. Grimshaw, M.K. Mohan, B.P. Nutley and M. Jarman (1999), 'Inhibition of protein prenylation by metabolites of limonene' *Biochemical Pharmacology*, 57(7), pp. 801–9.

26 T.H. Tsai, C.T. Huang, A.Y. Shum and C.F. Chen (1999), 'Simultaneous blood and biliary sampling of esculetin by microdialysis in the rat' *Life Sciences*, 65(16), pp. 1647–55.

27 Kohlert et al., 'Bioavailability and pharmacokinetics of natural volatile terpenes', p. 501.

28 A.A. Falk et al., 'Uptake, distribution and elimination of alpha-pinene', pp. 372–8.

29 A. Falk-Filipsson et al., 'D-limonene exposure to humans by inhalation', p. 84.

30 C. Kohlert et al., 'Bioavailability and pharmacokinetics of natural volatile terpenes', pp. 495–505.

31 R. Tisserand and T. Balacs (1995), *Essential Oil Safety: A Guide for Health Professionals*, Churchill Livingstone, Edinburgh, pp. 25, 46 (Table 5.2).

32 K.P. Svoboda et al., 'An investigation into drop sizes of essential oils using different dropper types', Table 3, p. 102.

33 R.J. Lambert, P.N. Skandamis, P.J. Coote, G.J. Nychas (2001), 'A study of the minimum inhibitory concentration and mode of action of oregano essential oil, thymol and carvacrol' *Journal of Applied Microbiology*, 91(3), pp. 453–62.

34 A.J. Hayes and B. Markovic (2002), 'Toxicity of Australian essential oil *Backhousia citriodora* (Lemon myrtle). Part 1. Antimicrobial activity and in vitro cytotoxicity' *Food Chemicals and Toxicology*, 40(4), pp. 535–43.

35 R. Tisserand and T. Balacs, *Essential Oil Safety*, p. 55.

36 H. Igimi, D. Watanabe, F. Yamamoto, S. Asakawa, K. Toraishi and H. Shimura (1992), 'A useful cholesterol solvent for the medical dissolution of gallstones' *Gastroenterologica Japonica*, 27(4), pp. 536–45.

37 R. Aeschbach, J. Loliger, B.C. Scott, A. Murcia, J. Butler, B. Halliwell and O.I. Aruoma (1994), 'Antioxidant actions of thymol, carvacrol, 6-gingerol, zingerone and hydroxytyrosol' *Food & Chemical Toxicology*, 32(1), pp. 31–6.

38 R. Tisserand and T. Balacs, *Essential Oil Safety*, p. 38.

39 M.J. Yin, Y. Yamamoto and R.B. Gaynor (1998), 'The anti-inflammatory agents aspirin and salicylate inhibit the activity of I(kappa)Bkinase-beta' *Nature*, November 5, 396(6706), pp. 77–80.

40 G. Lyss, T.J. Schmidt, I. Merfort and H.L. Pahl (1997), 'Helenalin, an anti-inflammatory sesquiterpene lactone from Arnica, selectively inhibits transcription factor NF-kappaB' *Biological Chemistry*, 378(9), pp. 951–61.

41 G.J. Tortora and S.R. Grabowski (1993), *Principles of Anatomy and Physiology*, 7th edn, HarperCollins, pp. 693–6.

42 H.P. Rang and M.M. Dale, *Pharmacology*, p. 197.

43 U.R. Juergens, M. Stober, L. Schmidt-Schilling, T. Kleuver and H. Vetter (1998), 'Anti-inflammatory effects of eucalyptol (1,8-cineole) in bronchial asthma: inhibition of arachidonic acid metabolism in human blood monocytes ex vivo' *European Journal of Medical Research*, 3(9), pp. 407–12.

44 N. Beuscher, M. Kietzmann, E. Bien and P. Champeroux (1998), 'Interference of myrtol standardized with inflammatory and allergic mediators' *Arzneimittel-Forschung*, 48(10), pp. 985–9.

45 H. Safayhi, S.E. Boden, S. Schweizer and H.P.T. Ammon (2000), 'Concentration-dependent potentiating and inhibitory effects of Boswellia extracts on 5-lipoxygenase product formation in stimulated PMNL' *Planta Medica*, 66(2), pp. 110–13.

46 H.P. Rang and M.M. Dale, *Pharmacology*, pp. 19–22, 265–6.

47 M. Lis-Balchin, S. Hart and E. Simpson (2001), 'Buchu (*Agathosma betulina* and *A. crenulata*, Rutaceae) essential oils: their pharmacological action on guinea-pig ileum and antimicrobial activity on microorganisms' *Journal of Pharmacy & Pharmacology*, 53(4), pp. 579–82.

48 L.P. Qin, H. Wu and G.H. Zhou (1993), 'Effects of total coumarins, essential oil and water extracts of *Cnidium monnieri* on plasma prostaglandin and cyclic nucleotide in the rats of

kidney-yang insufficiency' *Zhongguo Zhong Xi Yi Jie He Za Zhi Zhongguo Zhongxiyi Jiehe Zazhi*, 13(2), pp. 100–1.

49 R. Tisserand and T. Balacs, *Essential Oil Safety*, p. 38.

50 P.L. Crowell (1999), 'Prevention and therapy of cancer by dietary monoterpenes' *Journal of Nutrition*, 129(3), pp. 775S–8S.

51 W. Rabl, F. Katzgraber and M. Steinlechner (1997), 'Camphor Ingestion for Abortion (Case Report)' *Forensic Science International*, 89(1–2), pp. 137–40.

52 I.F. Delgado, A. Nogueira, C.A.M. Souza, A.M.N. Costa, L.H. Figueiredo, A.P. Mattos, I. Chahoud and F.J.R. Paumgarten (1993), 'Perinatal and postnatal developmental toxicity of beta-myrcene in the rat' *Food & Chemical Toxicology*, 31(9), pp. 623–8.

53 P.L. Crowell (1999), 'Prevention and therapy of cancer by dietary monoterpenes' *Journal of Nutrition*, 129(3), pp. 775S–8S.

54 N.S. Perry, P.J. Houghton, A. Theobald, P. Jenner and E.K. Perry (2000), 'In-vitro inhibition of human erythrocyte acetylcholinesterase by *Salvia lavandulaefolia* essential oil and constituent terpenes' *J. Pharm. Pharmacol.*, 52(7), pp. 895–902.

55 N.S. Perry, P.J. Houghton, P. Jennes, A. Keith and E.K. Perry (2002), '*Salvia lavandulaefolia* essential oil inhibits cholinesterase in vivo' *Phytomedicine*, 9(1), pp. 48–51.

56 H.P. Rang and M.M. Dale, *Pharmacology*, p. 141.

57 C. Fairlie, L. Baldwin, L. Vear and C. Rogers (1998), 'Bath PUVA: An Effective Treatment for Psoriasis' *Dermatology Nursing*, 10(4), pp. 285–9.

58 D. Engelstein, J. Shmueli, S. Bruhis, C. Servadio and A. Abramovici (1996), 'Citral and testosterone interactions in inducing benign and atypical prostatic hyperplasia in rats' *Comparative Biochemistry & Physiology Part C: Toxicology & Pharmacology*, 115(2), pp. 169–77.

59 R. Tisserand and T. Balacs, *Essential Oil Safety*, pp. 107–9.

60 H.P. Rang and M.M. Dale, *Pharmacology*, pp. 248–9.

61 G. Reid, A. Babes and F. Pluteanu (2002), 'A cold- and menthol-activated current in rat dorsal root ganglion neurones: properties and role in cold transduction' *Journal of Physiology–London*, 545(2), pp. 595–614.

62 N. Galeotti et al., 'Local anaesthetic activity of (+)- and (–)-menthol', pp. 174–6.

63 S. Seltzer (1992), 'Biologic properties of eugenol and zinc oxide-eugenol' *Oral Surgery Oral Medicine Oral Pathology Oral Radiology and Endodontics*, 73, pp. 729–37.

64 C. Ghelardini, N. Galeotti and G. Mazzanti (2001), 'Local anaesthetic activity of monoterpenes and phenylpropanes of essential
oils' *Planta Medica*, 67(6), pp. 564–6.

65 U.C. Luft, R. Bychkov, M. Gollasch, V. Gross, J.B. Roullet,
D.A. McCarron, C. Ried, F. Hofmann, Y. Yagil, C. Yagil,
H. Haller and F.C. Luft (1999), 'Farnesol blocks the L-type Ca^{2+}
channel by targeting the alpha 1C subunit' *Arteriosclerosis
Thrombosis and Vascular Biology*, Apr, 19(4), pp. 959–66.

66 P.M. Zygmunt, B. Larsson, O. Sterner, E. Vinge and E.D.
Hogestatt (1993), 'Calcium antagonistic properties of the
sesquiterpene T-cadinol and related substances: structure-activity
studies' *Pharmacology & Toxicology*, 73(1), pp. 3–9.

67 H.P. Rang and M.M. Dale, *Pharmacology*, p. 250.

68 J.M. Hills and P.I. Aaronson (1991), 'The mechanism of action of
peppermint oil on gastrointestinal smooth muscle. An analysis
using patch clamp electrophysiology and isolated tissue pharmacology in rabbit and guinea pig' *Gastroenterology*, Jul, 101(1),
pp. 55–65.

69 M.H. Pittler and E. Ernst (1998), 'Peppermint oil for irritable
bowel syndrome: a critical review and metaanalysis' *American
Journal of Gastroenterology*, 93(7), pp. 1131–5.

70 A.A. Albuquerque, A.L. Sorenson and J.H. Leal-Cardoso (1995),
'Effects of essential oil of *Croton zehntneri*, and of anethole and
estragole on skeletal muscles' *J. Ethnopharmacol.*, Nov 17, 49(1),
pp. 41–9.

71 H.P. Rang and M.M. Dale (1987), *Pharmacology*, pp. 19–23,
189–91.

72 A.N. Coelhodesouza, E.L. Barata, P.J.C. Magalhaes, C.C. Lima
and J.H. Lealcardoso (1997), 'Effects of the essential oil of
Croton zehntneri, and its constituent estragole on intestinal
smooth muscle' *Phytotherapy Research*, 11(4), pp. 299–304.

73 F.A. Santos and V.S.N. Rao (2002), 'Possible role of mast cells in
cineole-induced scratching behavior in mice' *Food and Chemical
Toxicology*, 40(10), pp. 1453–7.

74 A.J. Pinching (1977), 'Clinical testing of olfaction reassessed'
Brain, 100(2), pp. 377–88.

75 M.J. Howes, P. J. Houghton, D.J. Barlow, V.J. Pocock and S.R.
Milligan (2002), 'Assessment of estrogenic activity in some
common essential oil constituents' *Journal of Pharmacy & Pharmacology*, 54(11), pp. 1521–8.

76 N.S.L. Perry, P.J. Houghton, J. Sampson, A.E. Theobold, S. Hart,

M. Lis-Balchin, J.R.S. Hoult, P. Evans, P. Jenner, S. Milligan and E.K. Perry (2001), 'In-vitro activity of *Salvia lavandulaefolia* (Spanish sage) relevant to treatment of Alzheimer's disease' *Journal of Pharmacy & Pharmacology*, 53(10), pp. 1347–56.

77 G. Schmaus and K.H. Kubeczka (1985), 'The influence of isolation conditions on the composition of essential oils containing linalool and linalyl acetate', in A. Baerheim Svendsen and J.J.C. Scheffer (eds), *Essential Oils and Aromatic Plants—Proceedings of the 15th International Symposium on Essential Oils*, Noordwijkerhout, The Netherlands, July 19–21, 1984, Martinus Nijhoff/Dr W. Junk, Dordrecht, pp. 127–33.

78 K.M. Hold, N.S. Sirisoma, T. Ikeda, T. Narahashi and J.E. Casida (2000), 'Alpha-thujone (the active component of absinthe): gamma-aminobutyric acid type A receptor modulation and metabolic detoxification' *Proc. Natl. Acad. Sci. USA*, Apr 11, 97(8), pp. 3826–31.

79 L.F.S. Brum, E. Elisabetsky and D. Souza (2001), 'Effects of linalool on binding of [H-3]MK801 (NMDA antagonist) and [H-3]muscimol (GABAA agonist) to mouse cortical membranes' *Phytotherapy Research*, 15(5), pp. 422–5.

80 N. Perry, G. Court, N. Bidet, J. Court and E.K. Perry (1996), 'European herbs with cholinergic activities: potential in dementia therapy' *International Journal of Geriatric Psychiatry*, 11, pp. 1063–9.

81 H.P. Rang and M.M. Dale, *Pharmacology*, pp. 113–4.

82 B. Steinheider (1999), 'Environmental odours and somatic complaints' *Zentralbl. Hyg. Umweltmed.*, 202(2–4), pp. 101–19.

83 A. Seeber, C. van Thriel, K. Haumann, E. Kiesswetter, M. Blaszkewicz and K. Golka (2002), 'Psychological reactions related to chemosensory irritation' *Int. Arch. Occup. Environ. Health*, 75(5), pp. 314–25.

84 R. Baron (1990), 'Environmentally induced positive affect: its impact on self-efficacy, task performance, negotiation and conflict' *Journal of Applied Psychology*, (16), pp. 16–28.

85 T. Betts and L. Greenhill (pers. comm., 2002) 'Aromatherapy and Hypnosis in the management of epilepsy', Seizure Clinic & Epilepsy Liaison Service, Birmingham University, Queen Elizabeth Psychiatric Hospital, Birmingham, B15 2QZ UK, UK.

86 M. Moss, J. Cook, K. Wesnes, P. Duckett (pers. comm., 2002), 'Aroma of Rosemary and Lavender essential oils differentially affect cognition and mood in healthy adults', Human Cognitive

Neuroscience Unit, Division of Psychology, University of Northumbria, Newcastle-upon-Tyne, UK.

87 L.M. Levy, R.I. Henkin, C.S. Lin and A. Finley (1999), 'Rapid imaging of olfaction by functional MRI (fMRI): identification of presence and type of hyposmia' *J. Comput. Assist. Tomogr.*, 23(5), pp. 767–75.

Chapter 6

1 W.A. Poucher (1959), *Perfumes, Cosmetics and Soaps*, Chapman Hall, London; *Perfumer & Flavorist*, Allured Publishing Corporation, Carol Stream, IL, USA.

2 G. Schmaus and K.H. Kubeczka (1985), 'The influence of isolation conditions on the composition of essential oils containing linalool and linalyl acetate', in *Essential Oils and Aromatic Plants—Proceedings of the 15th International Symposium on Essential Oils*, pp. 127–33.

3 D.A. Moyler and H.B. Heath (1988), 'Liquid carbon dioxide extraction of essential oils', in *Flavors and Fragrances: A World Perspective. Proceedings of the 10th International Congress of Essential Oils, Fragrances and Flavors*, Washington DC, USA, 16–20 November 1986, Elsevier Science Publishers BV, Amsterdam, The Netherlands, pp. 41–63.

4 ibid., p. 54.

5 Website of Instituto Superior Técnico, Technical University of Lisbon, Portugal, http://alfa.ist.utl.pt/~fidel/flaves/intro.html, go to section 5 on Extraction.

6 D. Gerard, K.-W. Quirin and E. Schwarz (1995), 'CO2-Extracts from Rosemary and Sage: Effective natural antioxidants' *International Food Marketing & Technology*, October, pp. 46–55.

Chapter 7

1 M.H. Boelens, H. Boelens and L.J. van Gemert (1993), 'Sensory properties of optical isomers' *Perfumer & Flavorist*, vol. 18, no. 6, pp. 2–16.

Chapter 8

1 One of the more useful references for common essential oil components was volume 2 of *Natural Products of Woody Plants*, edited by J.W. Rowe and published by Springer-Verlag, Berlin, 1989.

Appendix

1 F. Chialva, F. Monguzzi, P. Manitto and A. Akgul (1993), 'Essential oil constituents of *Trachyspermum copticum* fruits' *Journal of Essential Oil Research*, Jan/Feb, vol. 5, pp. 105–6.

2 S.R. Srinivas (1986), 'Composition of angelica root oil' in S.R. Srinivas (ed.), *Atlas of Essential Oils*, Bronx, NY.

3 R. Tabacchi, J. Garnero and R. Buil (1974), 'Contribution a l'étude de la composition de l'huile essentielle de fruits d'anise de Turque' *Rivista Italiana*, vol. 56, pp. 683–97.

4 J.Q. Cu, F. Perineau and G. Goepfort (1990), 'GC/MS analysis of star anise oil' *Journal of Essential Oil Research*, Mar/Apr, vol. 2, pp. 91–2.

5 J.C. Chalchat, R.P. Garry, A. Michet and M. Gorunovic (1991), 'Essential oils of *Artemisia annua* from Yugoslavia' *Rivista Italiana EPPOS* (Special Issue), pp. 471–6.

6 A.C. Ronse and H.L. De Pooter (1990), 'Essential oil production by Belgian *Artemisia alba* (Turra) before and after micropropagation' *Journal of Essential Oil Research*, Sept/Oct, vol. 2, pp. 237–42.

7 G. Vernin, J. Metzger, D. Fraisse, K.-N. Suon and C. Scharff (1984), 'Analysis of basil oils by GC-MS Data Bank' *Perfumer & Flavorist*, Oct/Nov, vol. 9, pp. 71–86.

8 M.M. Carmo, E.J. Raposo, F. Venancio, S. Frazao and R. Seabra (1990), 'The essential oil of *Ocimum basilicum* L. from Portugal' *Journal of Essential Oil Research*, Sept/Oct, vol. 2, pp. 263–4.

9 D. McHale, W.A. Laurie and M.A. Woof (1977), 'Composition of West Indian bay oils' *Food Chemistry*, vol. 2, pp. 19–25.

10 G. Dugo, A. Cotroneo, A. Verzera, M.G. Donato, R. del Duce and G. Licandro (1989), 'Genuineness characters of the Calabrian bergamot essential oil', in *Proceedings of 11th International Congress of Essential Oils, Fragrances and Flavours*, 12–16 Nov, New Delhi, India, vol. 4, pp. 245–64.

11 L.J. van Gemert, L.M. Nijssen and H. Maarse (1983), 'Kwaliteitscriteria voor de kruiden Nootmuskaat en Zwarte Peper', TNO-CIVO Food Analysis Institute, private communication (to Boelens).

12 E. Klein and W. Rojan (1968), 'The most important constituents of buchu leaf oil' *Dragoco Report*, vol. 15, pp. 3, 4.

13 O. Motl, J. Hodacova and K. Ubik (1990), 'Composition of Vietnamese cajeput essential oil' *Flavor & Fragrance Journal*, vol. 5, pp. 39–42.

14 U.N. Senanayake (1977), 'The nature, description and biosynthesis of volatiles in *Cinnamomum* ssp.' Ph.D. thesis, University of New South Wales, Australia.

15 A.M. Janssen (1989), 'Antimicrobial activities of essential oils: a pharmacognostical study', Ph.D. thesis, Rijksuniversiteit, Leiden, The Netherlands.

16 J.C. Pieribattesi, J. Smadja and J.M. Mondon (1988), 'Composition of the essential oil of cardamom from Reunion', in B.M. Lawrence, B.D. Mookherjee and B.J. Willis (eds), *Flavors and Fragrances: A World Perspective, Proceedings of the 10th International Congress of Essential Oils, Fragrance and Flavors*, Washington DC, USA, 16–20 Nov 1986, Elsevier Science Publishers BV, Amsterdam, pp. 697–706.

17 H.L. De Pooter, B. Nicolai, J. de Laet, L.F. de Buyck, N.M. Schamp and P. Goetghebens (1988), 'The essential oils of five *Nepeta* species: a preliminary evaluation of the IR use in chemotaxonomy by cluster analysis' *Flavour & Fragrance Journal*, 3(4), pp. 155–9.

18 D.Z. Simon and Beliveau (1987), 'Cedarleaf oil (*Thuja occidentalis*): extracted by hydro diffusion and steam distillation' *International Journal of Crude Drug Research*, vol. 25, pp. 4–6.

19 B.M. Lawrence (1980), 'Chemical composition of oils of cedarwood from Texas and Virginia' *Perfumer & Flavorist*, June/July, 5(3), p. 63.

20 ibid.

21 A.L. Tsutuslova and B.A. Antonova (1984), 'Analysis of Bulgarian daisy oil' *Maslo-Zhir. Prom. St.*, vol. 11, pp. 23, 24.

22 A. Hasebe and T. Oomura (1989), 'The constituents of essential oils from *Anthemis nobilis*' *Koryo*, vol. 161, pp. 93–101.

23 R.O.B. Wijesekera, A.L. Jayewardene, L.S. Rajapakse and K.H. Fonseka (1974), 'Volatile constituents of leaf, stem and root oils of Cinnamon' *Journal of Science, Food and Agriculture*, vol. 25, pp. 1211–20.

24 ibid.

25 R.O.B. Wijesekera (1973), 'The chemical composition and analysis of citronella oil' *Journal of the National Science Council (Sri Lanka)*, vol. 1, pp. 67–81.

26 K. Bruns, E. Heinrich and I. Pagel (1981), 'Citronellaoel: Untersuchung von Handels—und Hybridoelen verschiedener Provenenz', in K.H. Kubeczka and G. Thieme (eds), *Vorkommen und Analytik aetherischer Ole*, Band 2, Verlag Publishers, Stuttgart.

27 B.M. Lawrence (1990), 'Comparative chemical composition of commercial clary sage oils' *Perfumer & Flavorist*, July/Aug, vol. 15, p. 71.

28 E.M. Gaydou and R.P. Randriamiharisoa (1987), 'Multidimensional analysis of GC data: application to differentiation of clove bud and clove stem essential oil from Madagascar' *Perfumer & Flavorist*, 12(5), pp. 45–51.

29 M. Milos and A. Radonic (1996), 'Essential oil and glycosidically bound volatile compounds from Croatian *Cupressus sempervirens* L.' *Acta Pharm.*, 46, pp. 309–14.

30 B.M. Lawrence (1980), 'The composition of elemi oil' *Perfumer & Flavorist*, Feb/March, 5(1), p. 57.

31 M.H. Boelens (1985), 'Essential oils and aroma chemicals from *Eucalyptus globulus* (Labil.)' *Perfumer & Flavorist*, 9(6), pp. 1–14.

32 A. Akgul (1986), 'Studies on the essential oils from Turkish fennel seeds', in E.J. Brunke (ed.), *Progress in Essential Oils (Proceedings of the International Symposium on Essential Oils)*, Walter de Gruyter, Berlin, pp. 487–9.

33 R.S. Hunt and E. von Rudloff (1974), 'Chemosystematic studies in the Genus *Abies*: 1. Leaf and twig oil analysis of alpine and balsam firs (average of 10 analyses)' *Canadian Journal of Botany*, vol. 52, pp. 477–87.

34 G. Vernin (1989), 'GC/MS data bank analysis of the essential oils from *Boswellia frereana* Birdw. and *Boswellia carterii* Birdw.', in G. Charalambous (ed.), *Flavors and Off-Flavors*, Elsevier Science Publishers, Amsterdam, pp. 511–42.

35 S.M. Abdel Wahab, E.A. Abontabl, S.M. El-Zalabami, H.A. Fouad, H.L. de Pooter and B. El-Fallaha (1987), 'The essential oil of olibanum' *Planta Medica*, 53(4), pp. 382–4.

36 G. Vernin, J. Metzger, D. Fraisse and C. Sharf (1983), 'Etude des huiles essentielles par CG-SM-Banque SPECMA: essences de geranium (bourbon)' *Parfumerie, Cosmetiques et Aromes*, vol. 52, pp. 51–61.

37 C.C. Chen and C.T. Ho (1989), 'Volatile compounds in ginger oil generated by thermal treatment (steamdistilled oil)', chapter 34 in T.H. Parliament, R.J. McGorrin and C.T. Ho (eds), *Thermal Generation of Aromas*, American Chemical Society, Washington DC, pp. 366–75.

38 M.H. Boelens (1991), 'Critical review on the chemical composition of citrus oils (normal Israeli grapefruit oil 1988)' *Perfumer & Flavorist*, March/April, 16, pp. 17–34.

39 B. Lawrence (1978), 'The composition of ho leaf oil' *Perfumer & Flavorist*, pp. 32–5.

40 D. Joulain and M. Ragault (1976), 'Sur quelques nouveaux constituents de l' huile essentielle d'*Hyssopus*' *Rivista Italiana EPPOS*, vol. 58, pp. 129–31, 479–85.

41 P. Weyerstahl et al. (1986), 'Isolation and synthesis of compounds from the essential oil of *Helichrysum italicum*', in E.J. Brunke (ed.), *Progress in Essential Oil Research (Proceedings of the International Symposium on Essential Oils)*, Walter de Gruyter, Berlin, pp. 177–95.

42 S.R. Srinivas (1986), 'Chemical composition of jasmine absolute', in S.R. Srinivas (ed.), *Atlas of Essential Oils*, Bronx, New York, pp. 1016–23.

43 R. Kaiser (1988), 'New volatile constituents of *Jasminum sambac* L. Aiton', in B.M. Lawrence, B.D. Mookherjee and B.J. Willis (eds), *Flavors and Fragrances: A World Perspective. Proceedings of the 10th International Congress of Essential Oils, Fragrance and Flavors*, Washington DC, USA, 16–20 Nov 1986, Elsevier Science Publishers BV, Amsterdam, pp. 669–95.

44 G. Bonaga and G.C. Galetti (1985), 'Analysis of volatile components in juniper oil by high resolution gas chromatography and combined gas chromatography/mass spectrometry' *Analytical Chemistry*, vol. 75, pp. 131–6.

45 M.A. Webb (2002), *Bush Sense*, Griffin Press, Adelaide, p. 61.

46 A. Zola and J.P. Le Vanda (1979), 'Le lavandin grosso' *Parfumerie, Cosmétiques et Aromes*, Jan/Feb, vol. 25, pp. 60–2.

47 A. Le Ster, J. Touche, R. Linas and M. Derbesy (1986), 'Hauteprovence French lavender essential oil', *Proceedings of the 9th International Congress of Essential Oils*, Singapore, pp. 127–33.

48 T.J. de Pascual, T. Ovejero, J. Anaya, E. Caballero, J.M. Hernández and M.C. Caballero (1989), 'Chemical composition of the Spanish spike oil (lab-distilled sample)' *Planta Medica*, vol. 55, p. 398.

49 C. Cappello et al. (1981), 'Richerche chiche sulla composizione dei derivati agrumari Argentini Nota 1. Gli olii essenziali' *Essenze Derivati Agrumari*, pp. 229–33.

50 D. Pénoël and P. Franchomme (1990), '*L'Aromathérapie exactement*', Roger Jollois, Limoges, p. 347.

51 L. Haro and W.E. Faas (1985), 'Comparative study of essential oils of key and Persian limes (distilled Persian lime)' *Perfumer & Flavorist*, Oct/Nov, vol. 10, pp. 67–72.

52 M.H. Boelens and R. Jimenez (1989), 'The chemical composition of some Mediterranean citrus oils' *Journal of Essential Oil Research*, vol. 1, pp. 151–9.

53 R. Oberdieck (1981), 'Ein Beitrag zur Kenntnis und Analytik von Majoran (*Marjorana hortensis* Moench)' *Zeitschrift fur Naturforschung*, vol. 36, pp. 23–9.

54 B.M. Lawrence (1981), 'The essential oils of *Litsea cubeba*' *Perfumer & Flavorist*, June/July, vol. 6, p. 47.

55 G. Tittel, H. Wagner and R. Bos (1982), 'Über die chemische Zusammensetzung von Melissenoelen' *Planta Medica*, vol. 46, pp. 91–8.

56 O. Vostrowsky, T. Brosche, H. Ihm, R. Zintl and K. Knobloch (1981), 'Über die Komponenten des aetherischen Oelen aus *Artemisia absinthum* L.' *Zeitschrift fur Naturforschung*, vol. 36C, pp. 369–77.

57 R.A. Wilson and B.D. Mookherjee (1983), 'Characterization of aroma-donating components of Myrrh (headspace analysis)', Paper no. 400, *Proceedings of the 9th International Congress of Essential Oils*, Singapore.

58 M.H. Boelens and R. Jimenez (1991), 'The chemical composition of Spanish Myrtle leaf oils Part 1' *Journal of Essential Oil Research*, May/June, vol. 3, pp. 173–7.

59 M.H. Boelens and S.R. Jimenez (1988), 'Essential oils from Seville bitter orange (*Citrus aurantium* L. ssp. *amara*)' in B.M. Lawrence, B.D. Mookherjee and B.J. Willis (eds), *Flavors and Fragrances: A World Perspective. Proceedings of the 10th International Congress of Essential Oils, Fragrance and Flavors*, Washington DC, USA, 16–20 Nov 1986, Elsevier Science Publishers BV, Amsterdam, pp. 551–65.

60 P.A.R. Ramoanoelina, J.P. Bianchini, M. Andriantsiferana, J. Viano and E.M. Gaydon (1992), 'Chemical composition of niaouli essential oils from Madagascar (oil of chemotype 1)' *Journal of Essential Oil Research*, Nov/Dec, vol. 4, pp. 657–8.

61 Analytical Methods Committee (1984), 'Application of Gas-Liquid Chromatography to the analysis of essential oils. Part XI: Monographs for seven essential oils' *Analyst*, vol. 109, pp. 1343–60.

62 M. Koketsu, M.T. Magaihaes, V.C. Wilberg and M.G.R. Donaliso (1983), 'Oleos essenciais de frutos citricos cultivadas no Brazil' *Boletin do. Pesquisas AMBRAPA Centro Technologico Agricolto Alimentario*, vol. 7, p. 21.

63 U. Ravid and E. Putievsky (1986), 'Carvacrol and thymol chemotypes of east Mediterranean wild Labiatae herbs', in E.J. Brunke (ed.), *Progress in Essential Oil Research (Proceedings of the*

International Symposium on Essential Oils), Walter de Gruyter, Berlin, pp. 163–7.

64 R.P. Randriamihariosa and E.M. Gaydou (1987), 'Composition of palmarosa oil' *Journal of Agriculture and Food Chemistry*, vol. 35, pp. 62–6.

65 S.R. Srinivas (1986), 'Composition of patchouli oil' in S.R. Srinivas (ed.), *Atlas of Essential Oils*, Bronx, NY.

66 F.W. Hefendehl (1970), 'Betirage zur Biogenese aetherische Oele Zusammensetzung zweier Aetherischer Oele von *Mentha pulegium* L. (C.R. variety)' *Phytochemistry*, vol. 9, pp. 1985–95.

67 M.B. Embong, L. Steele, D. Hadziyev and S. Molnar (1977), 'Essential oils from herbs and spices grown in Alberta' *Journal d'Institute de Canadien Science de Technologie Alimentaire*, 10(4), pp. 247–56.

68 M.H. Boelens and S.R. Jimenez (1986), 'Essential oils from Seville bitter orange', in B.M. Lawrence, B.D. Mookherjee and B.J. Willis (eds), *Flavors and Fragrances: A World Perspective. Proceedings of the 10th International Congress of Essential Oils, Fragrance and Flavors*, Washington DC, USA, 16–20 Nov 1986, Elsevier Science Publishers BV, Amsterdam, pp. 551–65.

69 K.H. Kubeczka and W. Schultze (1987), 'Biology and chemistry of conifer oils' *Flavour and Fragrance Journal*, 2(4), pp. 137–48.

70 ibid.

71 ibid.

72 J. Garnero, G. Guichard and P. Buil (1976), 'L'huile essentielle et la concentre de rose de Turquie' *Parfumerie, Cosmétiques et Savons*, (8), pp. 33–46.

73 M.S. Karawya, F.M. Hashim and M.S. Hifnawy (1974), 'Oils of jasmin, rose and cassie of Egyptian origin' *Bulletin of the Faculty of Pharmacy, University of Cairo*, vol. 13, pp. 183–92.

74 M.H. Boelens (1985), 'The essential oils from *Rosmarinus officinalis* L.' *Perfumer & Flavorist*, Oct/Nov, vol. 10, pp. 21–37.

75 G. Fournier, J. Habib, A. Reguigui, F. Safta, S. Guetari and R. Chemli (1989), 'Etude de divers echantillons d'huile essentielle de Romarin de Tunisie' *Plantes Medicinales et Phytotherapie*, vol. 23, pp. 180–5.

76 K. Formacek and K.H. Kubeczka (1982), 'The chemical composition of a commercial bois-de-rose oil', in *Essential Oils Analysis by Capillary Chromatography and Carbon-13 NMR Spectroscopy*, J. Wiley & Sons, New York.

77 E.A. Aboutab, A.A. Elazzouny and F.J. Hammerschmidt (1988), 'The essential oil of *Ruta graveolens* L. growing in Egypt' *Pharmacological Science*, vol. 56, pp. 121–4.

78 G. Vernin and J. Metzger (1986), 'Analysis of sage oils by GC-MS Data Bank' *Perfumer & Flavorist*, 11(5), pp. 79–84.

79 E.J. Brunke and F.J. Hammerschmidt (1988), 'Constituents of East Indian sandalwood oil: an eighty year long "stability test" (concentration in fresh oil)' *Dragoco Report*, no. 4, pp. 107–33.

80 F. Chialva, P.A.P. Liddle, F. Ulian and P. de Smedt (1980), 'Indagine sulla composizione dell olio ess. di *Satureja hortensis* L. coltivata in Piemonte e confronto con altr di diversa origine' *Rivista Italiana*, vol. 62, pp. 297–300.

81 S. Kokkikini and D. Vokou (1989), '*Mentha spicata* (Lamiaceae) chemotypes grown wild in Greece' *Economic Botany*, 43(2), pp. 192–202.

82 H.L. de Pooter, J. Vermeesch and N.M. Schamp (1989), 'The essential oils of *Tanacetum vulgare* L. and *Tanacetum parthenium* L.' *Journal of Essential Oil Research*, 1(1), pp. 9–13.

83 A.O. Tucker and M.J. Maciarello (1987), 'Plant identification', in J.E. Simon and L. Grant (eds), *Proceedings of the First National Herb Growing and Marketing Conference*, Purdue University Press, West Lafayette IN, pp. 126–72.

84 L.R. Williams and V.N. Home (1989), 'Plantations of *Melaleuca alternifolia*: a revitalised Australian tea tree oil industry', in *Proceedings of the 11th International Congress of Essential Oils, Fragrances and Flavours*, 12–16 Nov, vol. 3, pp. 49–53.

85 R. Piccaglia and M. Marotti (1991), 'Composition of the essential oil of an Italian *Thymus vulgaris* L. ecotype' *Flavour and Fragrance Journal*, vol. 6, pp. 241–4.

86 J.H. Zwaving and R. Bos (1992), 'Analysis of the essential oils of five *Curcuma* species' *Flavour and Fragrance Journal*, v.7, pp. 19–22.

87 I. Klimes and D. Lamparsky (1976), 'Vanilla volatiles: a comprehensive analysis' *International Flavour and Food Additives*, vol. 7.

88 J. Garnero (1972), 'A survey of the vetiver oil components (percentages estimated)' *Rivista Italiana EPPOS*, vol. 54, p. 315.

89 C. Frey (1988), 'Detection of synthetic flavorant addition to some essential oils by selected ion monitoring GC/MS (estimated peak percentages)' in B.M. Lawrence, B.D. Mookherjee and B.J. Willis (eds), *Flavors and Fragrances: A World Perspective. Proceedings*

of the 10th International Congress of Essential Oils, Fragrance and Flavors, Washington DC, USA, 16–20 Nov 1986, Elsevier Science Publishers BV, Amsterdam.

90 A.J. Falk, L. Bauer and C.L. Bell (1974), 'The constituents of the essential oil of *Achillea millefolium*' *Llodia*, vol. 37, pp. 598–602.

91 E.M. Gaydou, R. Randriamiharisoa and J.P. Bianchini (1986), 'Composition of the essential oil of ylang-ylang (*Cananga odorata* Hook F. et Thomson)' *Journal of Agriculture and Food Chemistry*, vol. 34, pp. 481–7.

INDEX

purity 160
see also sesquiterpenols
Juniper oil 167
Juniperus virginiana 74
see also sesquiterpenols

ketones 84–9, 106
molecular structure 184
Khella oil 109
see also coumarins
kidney irritation from monoterpenes 57

lactones 52, 101–6
molecular structure 187
Lavender oil 14–15, 114, 116
anti-spasmodic effect 70
decreasing memory performance
154
effects of steam distillation on 164
toxic effects of 92
treatment for burns and stings 133
see also monoterpenols
lecithin 48
lemon balm 36
Lemon eucalyptus oil 82
see also aldehydes
Lemon Myrtle oil 125
Lemongrass oil 75, 142
mucous membrane irritation from
82–3
see also aldehydes; sesquiterpenols
limbic system 152
Lime oil 166
limonene 125, 137
linalool 14–15, 34, 67, 92, 116
therapeutic uses of 69–70, 145
through steam distillation 164
see also monoterpenols
linalool oxide 98
see also cyclic ethers
linalyl acetate 92, 116
through steam distillation 164
see also acids and esters
lipophilic cell membrane interactions
135–6
lipophilic environment 42
lipophilic substances 45
liver
cytochrome P450 enzymes 141
damage through phenols 78
enzyme activity reduced by lactones
104
glutathione reduction 137–8
liver toxicity potential through
aldehydes 83
cyclic ethers 100

oxides 100
phenyl methyl ethers 96
lung excretion 126
see also drugs
lymphoedema 109

malaria *see* anti-malarial effects
mass spectrometry 163, 175
see also quality control
maximum tolerated dose (MTD) 132–3
see also drugs
May Chang oil 82
see also aldehydes
Melaleuca alternifolia 68
anti-bacterial effects of 135
Melaleuca quinquinervia oil 75
see also sesquiterpenols
melanoma 84
Melissa officinalis 36
Melissa oil 83
see also aldehydes
memory effects of essential oils 143,
150–1, 154
Mentha piperita oil 98
see also cyclic ethers
menthofuran 98
see also cyclic ethers
menthol 67, 69, 116
see also monoterpenols
menthone 85
see also ketones
metabolism of drugs 124–5
see also drugs
methane 15, 17
molecule representation 13
methyl chavicol 94, 138
see also phenyl methyl ethers
methyl eugenol 138
see also phenyl methyl ethers
methyl salicylate 92
see also acids and esters
mevalonic acid biosynthetic pathway 29
micelles 49–50
molecular structure 13, 15, 50–2
aldehydes 80, 183
carbon atoms 6–7
coumarins 106, 187
cyclic ethers 98, 186
drawing 13–15
esters 185
ethers 186
furocoumarins 187
hydrogen atoms 6
ketones 84–5, 184
lactones 101–2, 187
monoterpenes 53, 179–80

Printed in the United States
by Baker & Taylor Publisher Services